Continental Drift and Plate Tectonics

William Glen
College of San Mateo

Charles E. Merrill Publishing Company
A Bell & Howell Company
Columbus, Ohio

Published by
Charles E. Merrill Publishing Company
A Bell & Howell Company
Columbus, Ohio 43216

International Standard Book Number: 0–675–08799–6

Library of Congress Card Catalog Number: 74–75644

2 3 4 5 6 7 8 9 10 — 80 79 78 77 76 75

Cover illustration by Lou Hackett

Printed in the United States of America

Contents

Introduction

The most profound questions in geology have always been the simple ones. What makes the ground tremble during earthquakes? How did the mountains rise? Why are coal deposits, which are formed in warm climates, found in the Arctic? What caused the deep trenches in the ocean floor? What is the origin of the long chains of submarine volcanoes? Why are the fossils of once deep-sea-living creatures found in the rocks of high mountains? A revolutionary theory of incredible breadth has finally been developed that suddenly makes sense of things all over the world which have defied explanation since the dawn of man.

A glance at the globe reveals that some of the continents can be fitted easily together like pieces of a jigsaw puzzle and others joined with just minor changes along their edges. Although this idea had been touched on repeatedly since the 1800s, it was not until Alfred Wegener, a young German meteorologist, wrote a major work in 1915 that a storm of controversy was unleashed and half a century of inquiry was undertaken by hundreds of scientists in various disciplines. This great effort has culminated in the modern concept of continental drift and its offspring, plate tectonics.

The concept of continental drift holds that the earth's crust is broken into about one-half dozen major segments or plates and perhaps a dozen smaller ones. These plates are in continual motion, riding over a semimolten zone beneath them at the rate of a few inches per year. The continents are merely cargo carried about on the moving plates, which have been in motion for perhaps most of earth history. As plates move, oceans are created and destroyed, continents are torn apart or moved together to form larger ones, and earthquakes are triggered at the boundaries as plates separate, collide, dive, and slide past each other along great faults.

The concept of moving plates already has moved in part beyond the realm of theory since it is beginning to serve as a tool for making predictions. It gives the seismologist new insights into possible control of earthquakes which result from the release of strain that accumulates as plates move over, under, and past each other. Knowing the direction and speed of plate movement may allow the artificial triggering of quakes at locations of mounting strain before the strain reaches destructive proportions.

In addition to the earthly evidence, the space age has provided another tool for measuring plate movement. Laser beams fired at the moon and reflected back can be timed to fix the exact position of the observer on the earth and thus may soon help to measure the rate and direction of plate movement.

Oil and mineral deposits were fragmented as continents drifted, such that by reconstructing the jigsaw puzzle, mineral deposits on one continent allow us to predict their occurrence elsewhere. For example, the diamond-bearing rocks of West Africa are comparable to those of northeastern South America because they were once connected.

The fossil and rock records are punctuated by episodes of widespread, profound change in physical conditions with accompanying extinctions of many animal and plant groups. Such times also mark the origin of other life groups. The causes of such periods of accelerated decline and extinction of some groups, and apparently sudden rise and expansion of others, has long been in question. They may mark the intervals during which continents came together or were torn apart as a result of plate movement, destroying major seaways, altering the patterns of ocean current circulation, and producing great changes in climate. Such changes on a worldwide scale destroy old environments but simultaneously open new areas to invading groups, providing impetus for accelerated evolution.

The theory of plate tectonics impinges on virtually all areas of earth science and earth history. It appears to be the great unifying concept that geology has awaited since its earliest beginnings. It may affect the earth sciences as greatly as did Darwinian evolution biology in the last century or Copernicus did astronomy in the sixteenth century. Its coming has thrust on us a radically new perspective of the earth, the implications of which may be even further reaching than we can yet envisage.

The concept of continents riding on moving plates of crust must surely be earth science's greatest synthesis; it is a major modern scientific revolution which has grown out of the recent interpretation of geological facts accumulating for centuries. The real beginning of this revolution, unlike most others, is in the memories of living men, and its end may lie with those yet unborn.

1

The Crust and Interior of the Earth

In the sixteenth century, Giordano Bruno first compared the crust of the earth to the skin of an apple which wrinkles as the drying apple shrinks. The notion of a shrinking earth was reinforced by the Englishman Adam Sedgewick, who in 1833 explained that mountains formed as the earth cooled and contracted. A number of notables in geology later wrote to expand this contraction idea in both Europe and America.

The geologist of the nineteenth century thus came to envision the earth as a hot molten mass encased in an ever-shrinking, thin crust of cooler rock. Volcanic eruptions, hot springs, and high temperatures in deep mines and wells attested to the earth's internal heat, and the outpouring of lava was visible proof that the earth's interior was liquid. It was reasoned that the heat must be leftover from the fiery ball stage through which the earth had passed. The earth had been cooling from the surface inward since that time. The crust on which we live was the thin layer which had solidified and which would become thicker with passing time as the earth continued to cool. Eventually, the earth would solidify completely, lose its store of heat, and become a dead planet. This was geology's former great unifying theory which explained our past and served to predict the future.

Lord Kelvin, a leading physicist of the last century, had even calculated the time required for the earth to cool to its present state. His estimate of 20–40 million years since solidification of the crust was based on an earth which began as a hot gaseous mass which condensed as it cooled. A crust then began to form and continued to thicken to the present time. Geologists were not happy with Kelvin's short estimate of an earth age which was greatly at odds with their own. They knew that the processes which shaped the earth's features were the same slow processes operating at the present time. The slow shaping of a valley by streams moving a few grains of sand at a time, the gradual retreat of the sea cliff before the battering waves, and the sculpting of countless earlier valleys and sea cliffs all require time, much more time than allotted by Kelvin.

But Kelvin was a powerful force in the science of the nineteenth century. Although few among even the most prestigious scientists of the day cared to challenge him openly, Charles Darwin had taken issue with an earlier estimate of Kelvin's which allowed the earth a maximum of 400,000,000 years. Darwin argued that even this span of time was insufficient to account for the evolutionary changes indicated by the fossil record.

In spite of Kelvin's insistence on a comparatively short earth history, geologists of the last century owe much to him for his furthering of the cooling earth theory. However erroneous this concept was, it still provided a unifying theme to help explain the otherwise unexplainable, such as mountains and their folded and broken rock beds, the rise and fall of the land, and outpourings of lava. As the earth slowly cooled and solidified from outside in, it shrank. This contraction in turn resulted in the buckling of the thin, solid skin, much as the skin of an apple wrinkles when the apple dries and shrinks. Wrinkles in the crust of the earth formed great mountains and valleys. When the stresses could not be accommodated by wrinkling, the crust cracked, and rocks slid past each other along the resulting cracks, or faults. If the cracks opened to the base of the crust, then the fiery molten rock beneath the crust might work its way upward and pour out on the surface. All of this formed a neat picture which explained much to the nineteenth-century geologist.

This unifying theory was abruptly destroyed around the turn of the twentieth century as the result of two advances in science, the discovery of radioactivity, and the invention of the modern seismograph. Suddenly, geology found itself without a unifying theory, without a ready explanation for the most important geologic phenomena. A major goal of geology since then has been the construction of a new unifying theory.

The first blow to the old theory was dealt by the discovery of radioactivity. In 1896, the Frenchman de Becquerel found that certain substances, which we now call **radioactive isotopes,** spontaneously emit different kinds of radiation. Rutherford and Soddy demonstrated a few years later (1902) that radioactive isotopes break down to form new elements in a process called **radioactive decay.** One consequence of the decay is the liberation of a small amount of heat. It was soon discovered that radioactive substances are widely distributed in the rocks, any specimen of which may contain several such isotopes. Now, geologists had a hitherto undreamed of fuel which could keep the internal fires going almost indefinitely. It was no longer necessary to believe that the earth was cooling and shrinking; decay of radioactive isotopes could keep the temperature of the earth on an even keel. Indeed, it was no longer necessary to postulate an originally molten earth. Even an earth which began as a cold mass was plausible, for the earth could have acquired its heat through radioactive decay. A crust wrinkling to fit a contracting molten interior now seemed like the fantasy it always had been.

Seismology as a Probing Tool

The final blow was dealt to the contraction theory by the invention of the modern seismograph and the subsequent revelation, based on studies of earthquake waves, that the earth is solid to a depth of 1800 miles. A seismograph is a simple instrument, the main part of which is a heavy weight suspended such that it tends to remain stationary or move slowly while everything around it shakes during an earthquake. (See figs. 1.1a and 1.1b.) A recorder translates the shakes as wiggles on a sheet of paper. The record of the wiggles is called a **seismogram.** Much more sensitive to vibrations than any human, a seismograph can detect an earthquake on the opposite side of the world.Some of the shock waves set off during a quake are recorded after traveling through the earth's interior. Because the waves are affected by the material through which they travel, they can be interpreted by an expert to give a picture of the inside of the earth.

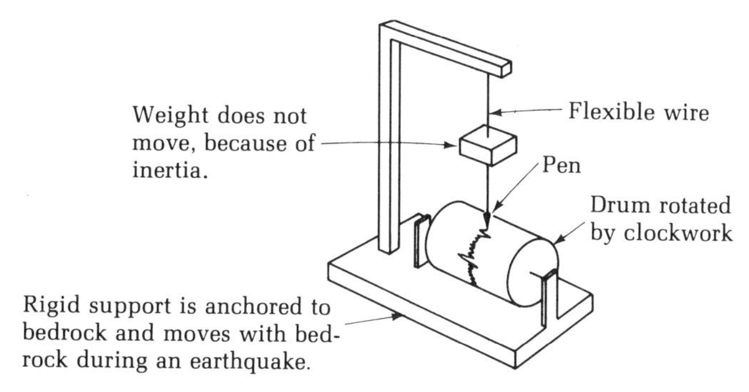

FIGURE 1.1a Seismograph. The base of the seismograph is firmly attached to bedrock so that all of the rigid and attached parts of the instrument vibrate and move with the earthquake waves. The weight suspended on the flexible wire tends to remain motionless because of inertia. Therefore, the pen attached to the weight records the relative movement between the rotating drum that moves with the quake and the motionless pen. (Reprinted, by permission, from Robert J. Foster, *General Geology*. 2nd ed. Columbus: Charles E. Merrill Publishing Co., 1973. Courtesy of the author.)

An earthquake occurs when rocks break in response to stresses within the earth and slide rapidly past each other along the fracture, or fault. The abrupt movement causes the ground-shaking we call an **earthquake.** The point at which the movement occurs is termed the **focus;** the **epicenter** is the point on the surface of the earth directly above the focus. For most earthquakes, the focus is within 40 miles (65 km) below the surface, but earthquakes have occurred at depths as great as 435 miles (700 km). We shall see later that the depth of quakes is closely related to major earth structures, and that earthquakes occur in definite zones.

At the moment an earthquake occurs, two distinctly different kinds of shock waves are generated, which then radiate outward in all directions from the focus. They are termed **body waves** because they travel through the interior of the earth. When the body waves reach the surface, they give rise to yet another type of complex wave, called a **surface wave** because it travels along the surface of the earth. Seismologists refer to the surface waves as **L waves.** (Actually, the L wave is a combination of two different kinds of waves.)

It is the body waves which give us the "x-ray image" of the earth's deep interior. Although generated simultaneously by the same rock movement, the

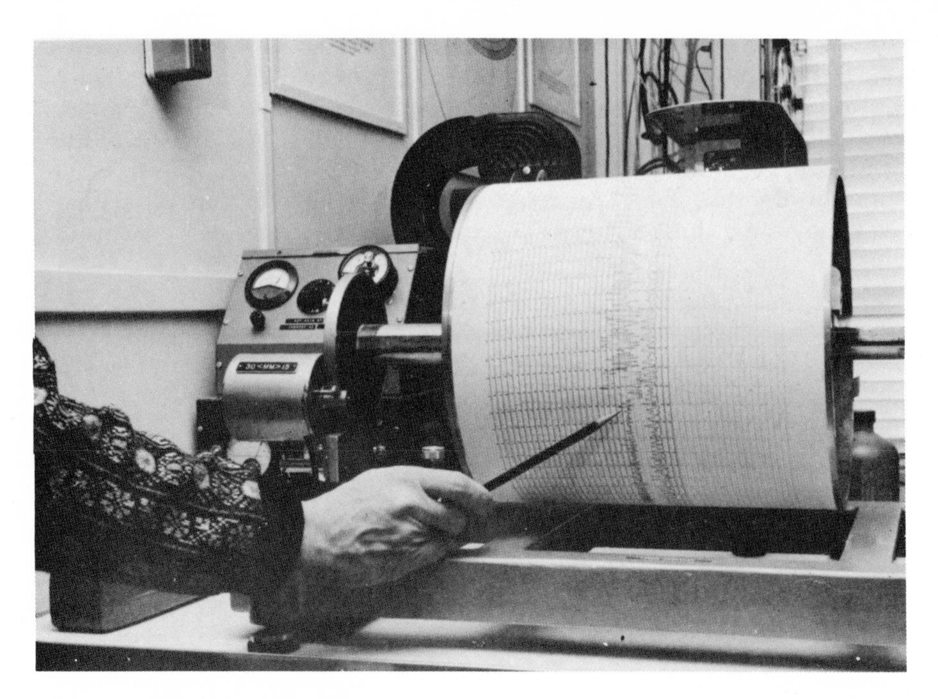

FIGURE 1.1b A Seismograph Designed to Sense and Record L Waves, or Surface Waves. (Photograph courtesy of Environmental Science Services Administration, U. S. Coast and Geodetic Survey.)

two body waves are propagated differently through the rocks. One wave is a **compressional wave;** as it passes through any material, each particle in the substance vibrates back and forth along the line of travel of the wave. Compressional waves can travel through anything — solid, liquid, or gas. Sound waves, for example, are compressional waves which travel through air (see fig. 1.2).

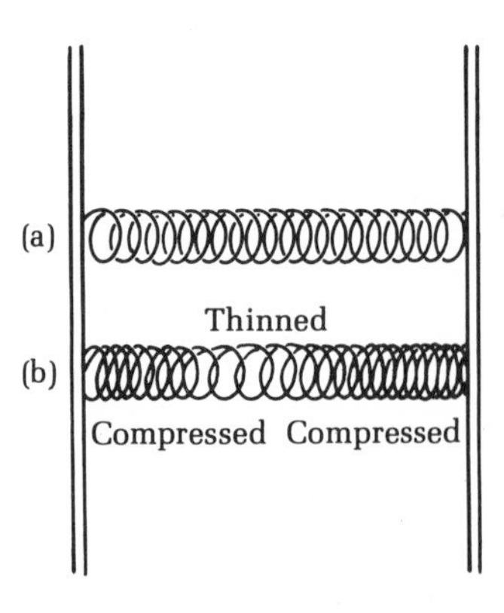

FIGURE 1.2 Propagation of a Compressional, or P, Wave. (a) The spring is stretched between supports. (b) A P wave moves along the spring, producing compressed and thinned regions. (From Foster, *General Geology*, 2nd ed.)

The second kind of body wave is a **transverse wave;** as it travels through a substance, particles of the substance vibrate at right angles to the line of travel of the wave. Unlike compressional waves, transverse waves can travel only through solids (see fig. 1.3).

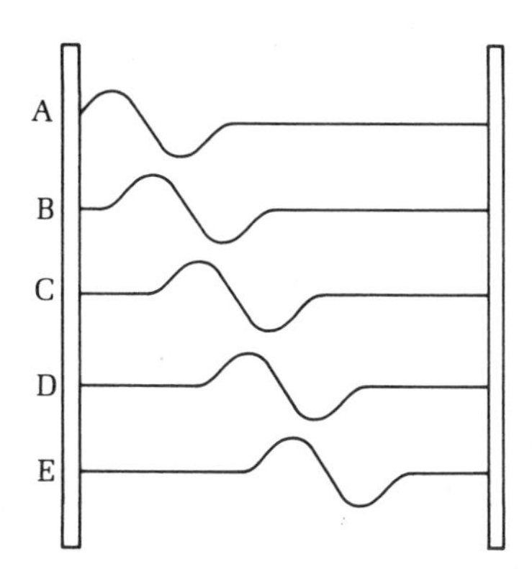

FIGURE 1.3 Propagation of a Transverse, or S, Wave along a Rope. The wave origin is shown at *A*; *B* through *E* show the rope in successive stages as the wave displaces it. (From Foster, *General Geology*, 2nd ed.)

These basic differences between the body waves result in different rates of travel. The faster compressional wave always arrives at a given point before the transverse wave, so it is recorded first on a seismogram. Therefore, the faster compressional wave is called the **primary wave,** and the transverse wave is called the secondary wave, abbreviated as the P wave and S wave, respectively. The L wave travels more slowly than the body wave and is always last to arrive and be recorded by a seismograph. When an earthquake registers at a seismograph within a few thousand miles of the epicenter, the instrument typically records the three waves in succession. First, the vibration due to the arrival of the P wave is recorded by the instrument. Then, somewhat later, another vibration due to the later arriving S wave is picked up, and finally, a third vibration registers the arrival of the L wave. Figure 1.4 shows how such a record, called a **seismogram,** might appear. In actual practice, the record is more complex than the illustration because waves which bounce back and forth inside the earth before arriving at the seismograph complicate the picture.

The speed of the body waves is very high — the P wave can make the journey

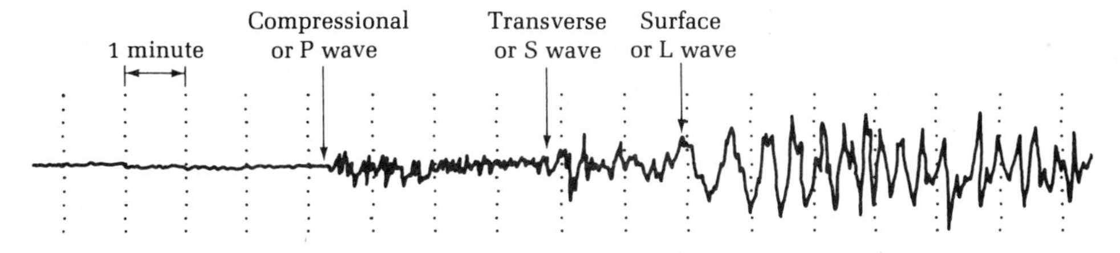

FIGURE 1.4 Seismogram. A written record from a seismograph shows the compressional (P) wave, the transverse (S) wave, and the surface (L) wave. (From Foster, *General Geology*, 2nd ed.)

to a point on the opposite side of the earth in about 20 minutes. The S wave travels about half as fast, although, as we will see, it never makes the trip completely through the earth. L waves take much longer to cover an equivalent distance, requiring several hours to travel halfway around the world. Seismologists use the difference in travel times to calculate the distance from a seismograph to the epicenter of an earthquake. After the difference in arrival times of the P and S waves at the seismograph is noted, to determine the distance to the point of origin of the earthquake is a simple matter. The greater the lag between the two waves, the greater the distance to their point of origin. It is much as though a fast passenger train traveling 100 miles per hour and a slower freight train traveling 50 miles per hour left a certain city together. But the faster passenger train will gain continually on the freight train. Imagine that you are standing on the platform of a station in a distant city and you observe that the passenger train arrives at the station one hour ahead of the freight train. If you know how fast they have been traveling, you can quickly calculate that the starting point was 100 miles up the track. If the freight had come in two hours after the passenger train, then their starting point must have been 200 miles away. Exactly the same principle is used to determine the distance to the focus of an earthquake. The time lag between the arrival of the P wave and the S wave at a seismograph tells the scientist the distance to the earthquake's point of origin. If at least three seismograph stations determine the distance to the focus from each station, the location of the epicenter can be obtained by drawing a circle around each station on a map such that the radius of each circle is equal to the distance to the earthquake's point of origin. The intersection of the three circles locates the epicenter (see fig. 1.5).

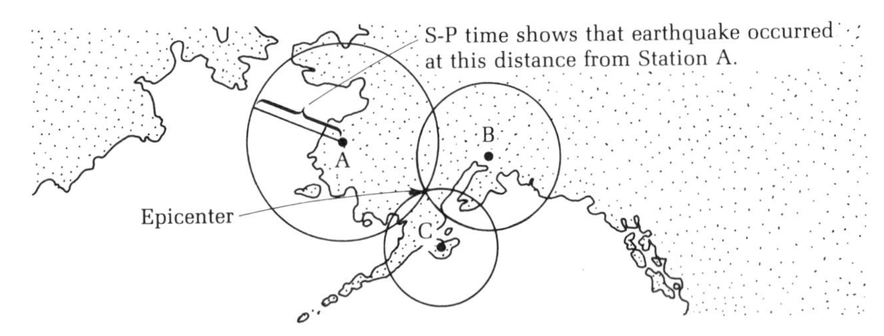

FIGURE 1.5 Epicenter. Seismograph stations are located at *A*, *B*, and *C*. The time lag between the arrival of the S and P waves at each station reveals that the earthquake occurred somewhere on the circle drawn around each station. The epicenter is located where the three circles intersect. (From Foster, *General Geology*, 2nd ed.)

Inferences of Interior Earth Structure

Over the years, seimologists laboriously have collected data relating speed of travel of body waves to depth penetrated by the waves. Those waves which are picked up at more distant seismographs have traversed the deeper layers of the earth. Waves picked up at seismographs near the focus have penetrated only the shallower depths. With great patience, scientists have determined the speed of P and S waves at different depths by timing then from a known focus. It is now known that both waves generally speed up as they penetrate deeper and deeper into the earth, but then at a depth of 1800 miles (2900 km) the speed of the P wave abruptly drops to about the same as its speed in the layers only a few miles below the surface, and the wave is sharply bent. The S wave disappears (see fig. 1.6). Such behavior of the P and S waves suggests to seismologists that the nature of the material in the earth's interior changes abruptly at a depth of 1800 miles. Failure of S waves to penetrate deeper is strong evidence that the material changes from solid to molten rock at that depth. Recall that S waves can travel only through solids. A change back from molten material to solid at a depth of 3160 miles (5150 km) is indicated by a small but sudden increase in the speed of the P wave. A study of earthquakes thus provides valuable clues to the earth's internal structure which can be obtained in no other way.

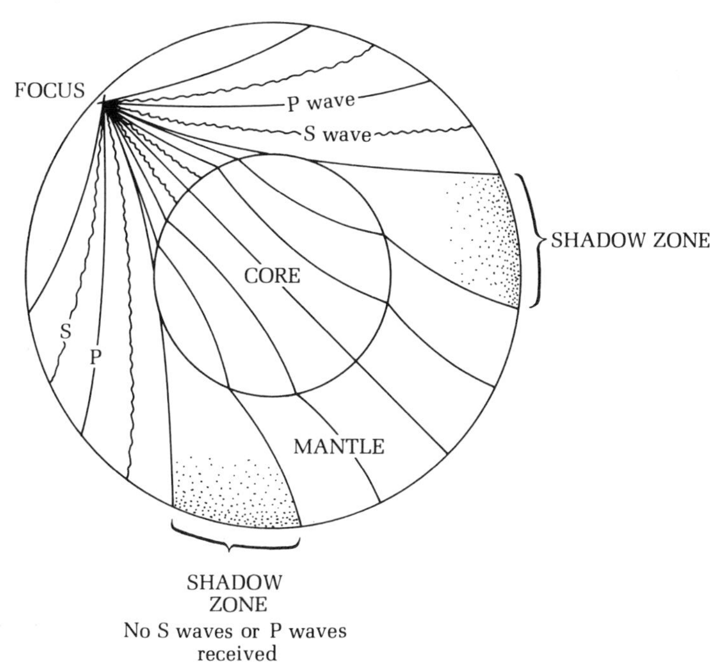

FIGURE 1.6 Effect of the Earth's Interior Zones on Earthquake Waves. The shadow zone is caused by the inward bending of compressional waves when they enter a zone in which they are slowed. The S waves have a transverse motion and thus are not able to penetrate the liquid core.

A picture of a layered earth has emerged gradually over the years (fig. 1.7). We now recognize a thin outer layer, called the **crust,** which is only 3 to 5 miles thick beneath the ocean basins and about 15 to 35 miles thick beneath the continents. Under the crust is the **mantle,** a much thicker layer which extends down to 1800 miles (2900 km) below the surface. (The mantle is treated in greater detail later.) In contrast to the overlying layers, the **outer core,** extending from the base of the mantle to a depth of 3160 miles (5150 km) is fluid. The innermost layer is the **inner core;** it lies below the outer core. Solid material seems to constitute this central portion of the earth.

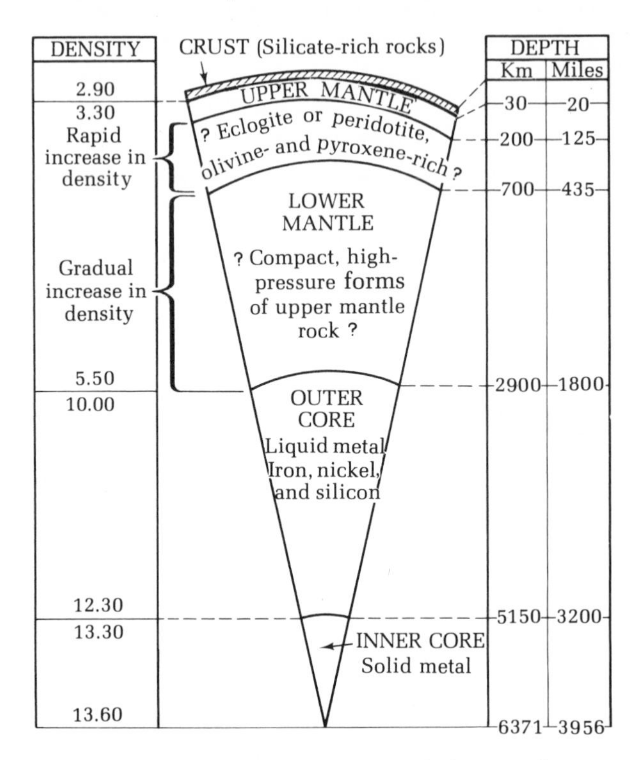

FIGURE 1.7 A Wedge Cross Section of the Earth's Internal Zones. The crust is exaggerated.

Changes in the state of matter from layer to layer influnce the behavior of the earthquake-generated body waves as they travel through those layers. The sharp bending of the P waves and the disappearance of the S waves at the 1800-mile depth result in the creation of a shadow zone. The shadow zone is a 3000-mile-wide belt which girdles the earth between a distance of 7000 and 10,000 miles from the point of origin of an earthquake. Seismographs within the shadow zone record no S wave and, as a result of sharp bending of the P waves at the outer core, no P wave. Beyond the shadow zone the P wave makes a reappearance, but again there is no S wave. Figure 1.6 illustrates the paths followed by P and S waves as they travel through the earth.The diagram shows the body waves traveling along a path which just barely enters the outer core. S disappears and P is bent so that it appears at a distance of 10,000 miles from the focus measured along the surface of the earth. The waves following the path which barely misses the outer core arrive at the surface 7000 miles from the epicenter.

The study of body waves provides much additional information about the physical properties of the earth's layers. For example, it tells us that the material inside the earth increases in density with depth. At the center of the earth the material is about five times as dense, or heavy, as the average for the crust. Much as we can learn about the earth's interior from earthquakes, however, a study of earthquake waves does not give us direct information about the chemical or mineral composition of the earth's interior. Except for the shallower parts of the crust, which are open to direct observation, we must base our picture of the actual composition on inference.

The Crust and Its Rocks

Direct observation and geophysical evidence reveal that the crust is composed largely of two kinds of rock. One is a dark-colored rock called **basalt,** in which the mineral grains are microscopic. The other is **granite,** a light-colored, coarse-grained rock in which the mineral grains can be seen easily. Both are **igneous** rocks, rocks which solidified from molten material. A familiar example of an igneous rock is obsidian, a glassy rock which forms by the quick cooling of lava at the surface during volcanic eruptions. Granite and basalt similarly form from molten rock, although granite forms when molten rock of the right chemical makeup solidifies slowly deep under the ground.

The crust of the ocean basins has the composition of basalt, which also comprises most of the volcanic islands of the world's oceans. Typical are the islands of Hawaii, which are composed of successive layers of lava flows piled one atop the other. Small crystals of the white mineral feldspar, or the black or brown mineral pyroxene, or the green, glassy-appearing olivine are embedded in a black matrix of microscopic crystals in a typical specimen of basalt. Other specimens of basalt may contain no visible crystals, consisting only of a dark-colored intergrowth of microscopic crystals. Analysis of basalt shows it to contain more of the elements iron, calcium, and magnesium than most other igneous rocks. Igneous rocks rich in those elements are commonly called **simatic** rocks. The oceanic crust is thus of simatic composition.

Most of the continental crust is made up of granitic rocks. Granitic rocks differ chemically from basalt or sima in that they are poor in iron, magnesium, and calcium but contain more of the elements sodium, potassium, and silicon. The term **sialic** is applied to igneous rocks which have such composition. Sial is thick in the continents but absent from the ocean basins. The thick sial layer of the continents was formerly assumed to be underlain by a globe-circling, sima layer which made up the ocean floor and was simply depressed but continuous under the continental sial layer. This simple picture has recently changed because it now appears that the granitic sial which makes up the bulk of the continents is not neatly layered above a distinctly separate, deep sial layer. Instead, granitic and basaltic rocks appear intermingled at depth under the continents and have confused the picture.

The other rocks present in the crust are the familiar **sedimentary** rocks such as sandstone, shale, limestone, and conglomerate. They are formed by the compaction and cementation of layers of sediment which are deposited mainly on the continental shelves and slopes bordering the continents under the shallower parts of the oceans and seas. Lesser amounts have been deposited in stream valleys and lakes on land. Although sedimentary rocks cover about three-fourths of the continental surface area and thus appear abundant and conspicuous, they make up only about 5% of the volume of the entire earth crust.

The third major rock family in addition to the igneous and sedimentary is the **metamorphic.** Metamorphic rocks are produced when pressure, heat, the introduction of new chemical substances, or a combination of these factors produces changes in composition, texture, or internal structure in any kind of preexisting rock. The common metamorphic rocks such as gneiss, schist, slate, or marble comprise only a very small fraction of the entire crust. The igneous rocks are by far the most important, for they comprise more than 90% of the entire crust, and thus receive most attention in discussions of major crustal and upper-mantle questions.

Man has never penetrated the mantle. Recent unsuccessful attempts have been made to drill through the ocean floor into the mantle. Until the day arrives when successful deep-drilling provides us with specimens of the mantle, we cannot be certain about its composition. However, all evidence points to a mantle having a chemical composition not too different from that of basalt. Likely candidates for the upper mantle are two rocks, eclogite and peridotite, both of which have the required physical properties. **Eclogite** is a pretty rock which consists of reddish-brown crystals of garnet surrounded by a bright-green mineral called **omphacite.** Both minerals form under high pressure, such as must exist in the mantle. Although the minerals found in eclogite are quite different from those in basalt, it is interesting to note that basalt and eclogite have the same chemical makeup. If the mantle consists of eclogite, it seems that basalt forms when eclogite melts; then, while fluid, it moves up through fractures in the earth to places of lower pressure. When the molten material solidifies, the garnet and omphacite cannot form because the pressure is too low; instead, the minerals of basalt crystallize. Therefore, basalt may be the low-pressure equivalent of eclogite. Laboratory experiments, however, have cast some doubt on eclogite as the main component of the mantle and as the main source of basalt.

Alternatively, the upper mantle may consist of a rock called **peridotite,** a simatic rock that is richer in iron and magnesium than is basalt. Olivine and the mineral pyroxene are the main constituents of peridotite. A strong argument in favor of a peridotite mantle is the similarity of peridotite to "stony" meteorites, in both chemical and mineral composition. Stony meteorites are similar in composition to earth rocks and are in sharp contrast to iron-nickel meteorites which are made largely of those two metals. The argument in favor of a peridotite upper mantle holds that the earth and meteorites are parts of the same solar system; the meteorites are debris leftover from the formation of the planets or are fragments of a larger body or bodies which shattered sometime in the past; therefore, it is reasonable to expect that similarities in composition should exist between earth and meteorites. At any rate, the physical properties of peridotite match those of the upper mantle as determined by geophysical methods. Whether the upper mantle is eclogite, peridotite, or something else, however, cannot be known with certainty until more data become available.

A rapid increase in the speed of earthquake body waves in a zone between 200 and 500 miles (325-800 km) deep reveals a transition in the mantle at those depths. It is thought that the very high pressures which prevail there are forcing the atoms of the mantle rocks to rearrange themselves into more compact groupings. New minerals form in which the atoms are packed more closely under the higher pressures. Below the transition zone, the lower mantle is made of the newer high-pressure minerals, although its overall chemical composition may be the same as that of the upper mantle.

Both the inner and outer core consist of material of high density, the composition of which must remain in the realm of speculation. In all likelihood, a sample of the core will never be obtained for study, for it seems impossible that a hole could ever be drilled to such a depth. Our best estimate is that the core is made of iron and nickel and has the same composition as iron-nickel meteorites. Again, the rationale is that the earth may have a composition similar to that of other bodies in the solar system. In addition, the physical properties of an iron-nickel mass under high pressure are thought to match those determined for the core. Theoretical considerations suggest that some lighter elements such as silicon and sulfur must also be present in the core, silicon because it is abundant in other rocks and sulfur because the known amount on earth is less than is to be expected on the basis of sulfur's abundance in the solar system.

In overview (see fig. 1.8), we may now say that the continents are underlain by a comparatively light granitic crust about 25 miles thick, and an oceanic crust only 5 miles thick composed of a denser basaltic rock. The boundary between the crust and mantle is called the **Mohorovičić discontinuity,** or the **Moho.** It had long been thought that the Moho was the truly important boundary along which large-scale movements of the crust took place in slipping over the uppermost mantle. This no longer appears so. It has recently been discovered that the speed of certain seismic waves suddenly decreases below a boundary about 40 miles (65 km) below the oceans and about 90 miles (140 km) under the continents. This suggests that there is an outer rigid shell about 40 to 90 miles thick which includes both the earth's crust and the uppermost part of the mantle — this shell is called the **lithosphere.**

The speed of travel of earthquake waves depends on the density and flow properties of the rock through which it passes. Waves move quickly in dense, rigid rock and more slowly in lighter, less dense material. Therefore, the zone beneath the rigid lithosphere in which the waves slow down markedly is now referred to as the **low-velocity zone** or **plastic layer.** This low-velocity part of the mantle is thought to be in a near-liquid state in which the rocks are close to their melting points. In fact, the rocks do melt in places inasmuch as volcanic lavas in Hawaii and New Zealand originate at a depth of about 40 miles, the depth at the top of the low-velocity zone beneath the oceans. This low-velocity or plastic zone is now referred to as part of the **asthenosphere** (weak layer). The asthenosphere appears weakest near its top, just below the rigid lithosphere, where it is thought to be able to undergo slow plastic flow of about a few inches per year. Don L. Anderson believes that the plastic layer of the asthenosphere extends from approximately 40 to 150 miles in depth. The asthenosphere becomes more viscous or rigid with depth, as indicated by the increase in earthquake wave speeds at about 250 to 400 miles.

Our new model of earth structure now has an outer rigid lithospheric shell 40 to 90 miles thick, composed of the crust and the rigid uppermost part of the mantle. The lithosphere overlies the asthenosphere, which includes an upper, weak, semimolten zone in which plastic flow appears to take place. This new

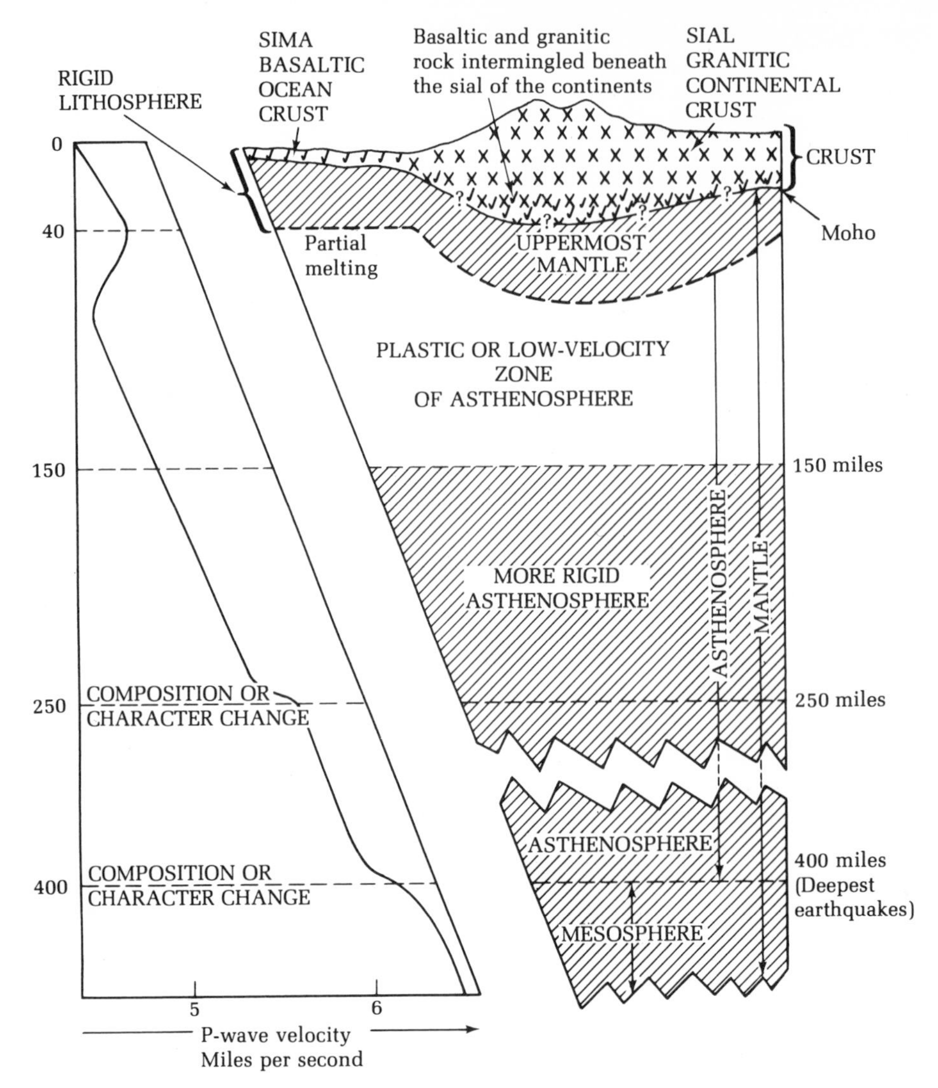

FIGURE 1.8 The Structure of the Crust and Mantle with P-Wave Velocities Shown. The thickness of the crust has been greatly exaggerated for illustration. Earthquake wave velocity depends greatly on the elasticity of rock. The rocks of the lithosphere, which are more rigid and elastic than those of the plastic zone of the asthenosphere below, transmit waves faster. The rocks of the mantle become plastic at a depth of about 40 to 90 miles — this depth is the top of the plastic, low-velocity zone of the asthenosphere. Deeper than 150 miles, the rocks again become more elastic; the waves speed up, showing marked increases at 250- and 400-mile depths within the asthenosphere. The mantle below the asthenosphere is the mesosphere. The boundary between them at 400 miles marks the zone of deepest known earthquakes. The boundary between the lithosphere and the asthenosphere is the important one along which movements have taken place to produce deformed crustal features throughout geologic time.

picture suggests that the physical boundary between the lithosphere and asthenosphere is the important one along which movements have taken place to produce the deformed crustal features which geologists have sought so long to explain. We shall again turn to evidence from the distribution of earthquakes which supports the very important idea of this relationship between lithosphere and asthenosphere in our chapter on plate tectonics.

The observation of the nineteenth-century geologists that the earth has an internal reservoir of heat was correct. It has been well known for many years that the deeper we penetrate the crust, the hotter are the rocks which we encounter. In deep mines the temperature can be unbearably hot without some means of cooling the air. Water in deep rock layers is warmer than underground water from shallow depths. Countless observations such as these have established firmly that temperature increases with depth. Measurements made at many points on earth reveal that on the average, the temperature increases 1°F for each 60 feet of depth, or about 90°F per mile. The rate of rise with depth is called the **geothermal gradient** (see figs. 1.9a and 1.9b). We know that the gradient cannot continue at that rate, however, for if we extend a gradient of 90°F per mile downward to the center of the earth, we arrive at a temperature of more than one-third of a million degrees, high enough to vaporize the earth. Clearly, the rate of temperature increase must level off at a fairly shallow depth. The temperature in the mantle is not high enough to melt the rocks, except locally in the low-velocity zone. Although the temperature of the mantle rocks probably is high enough to melt them if they were at the surface of the earth, the high pressure, which inhibits melting and thus raises the melting temperature of the rocks, helps keep the rocks in the solid state. If we were certain of the exact composition of the mantle and knew the effect of pressures on the melting temperatures of those materials, we could at least establish a maximum value for the temperature at a given depth. Our limited knowledge enables us to say only that the temperature of the mantle probably is less than 10,000°F.

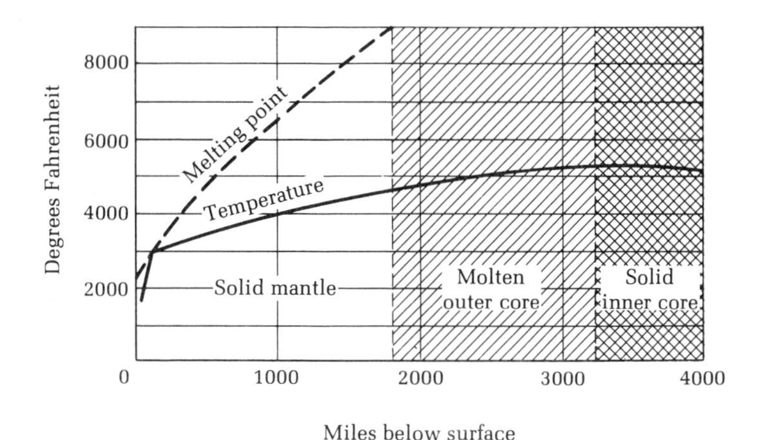

FIGURE 1.9a Estimated Temperatures in the Earth. Recent estimates of temperatures within the core are even higher than those indicated in the curve. (Reprinted, by permission, from John Verhoogen, "Temperatures Within the Earth." *American Scientist* 48, 1960.)

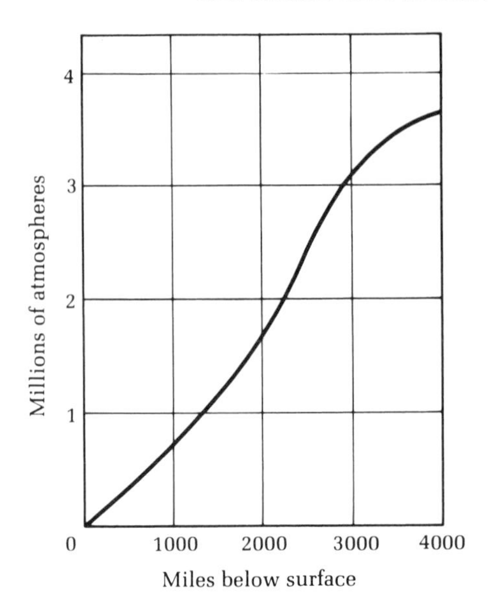

FIGURE 1.9b Estimated Pressures within the Earth. (From Verhoogen, *American Scientist* 48, 1960.)

Similarly, if we knew the exact composition of the core, we could make an accurate evaluation of the temperature at the boundary between the inner and outer core. The temperature there is at the melting point of the core. In all likelihood, the inner core is hotter than the outer core but is solid because the higher pressure at the greater depths counteracts the effect of the higher temperature and prevents the inner core from melting. About all that we can confidently say about the temperature of the core is that the core probably is not much hotter than the mantle.

Expanding Earth Theory

The idea of an expanding earth was first proposed in 1933 by O. C. Hilgenberg; he believed that both the mass and the size of the earth had grown. Shortly afterward, the astronomer J. K. E. Halm suggested that the earth's density has been decreasing from an original 9.13 to its present 5.5. In 1955, L. Egyed proposed that the earth's core is in an unstable state undergoing change to a state of lower density, resulting in an increase in volume. He further believed that the amount of water on the earth either has increased or at least has remained stable through earth history; he noted that earth expansion must be taking place in the ocean areas because the continental plates do not appear to allow this. Such expansion of the oceans must have proportionately reduced the area of the stable continents. Egyed used a series of elaborate paleogeographic maps showing the changing continents through time. S. W. Carey thought that his own jigsaw reconstructions for 200 million years ago would have been greatly improved by reducing the earth's diameter to about three-fourths of its present size, thus lending support to the notion of an expanding earth.

There are also some astronomical data which suggest an expanding globe; the earth's rate of rotation is slowing about two seconds per 100,000 years. This figure appears supported by the work of paleontologists who have been able to count daily growth rings in corals and other fossils which indicate that the earth had a year of about 400 days in the Devonian period 375 million years ago. In contrast, recent studies in paleomagnetism offer strong evidence that the earth could not have expanded by more than a few percent. The magnetic angles of inclination shown by magnetized rocks dating back as far as the Cambrian period 600 million years ago indicate that the earth had an essentially modern diameter at that time. Resolution of the question requires further data.

2

Geologic Time and Dating Rocks

Geologic Time Scale

Determining the age of the earth and discovering a means of calibrating earth history have been central questions in geology since its beginning, but it was not until 1762 that George Fuchsel of Germany equated various rock beds with time intervals and noted that earth history could be worked out by their study. He gave us the first notion of a geologic time scale.

To date a rock is to assign an age to it, to place it within a scale of time; its age may be expressed simply in relative terms, for example, bed A is older than bed B, but younger than bed C. The present geologic time scale was evolved by compiling rock beds formed during different geologic intervals at different localities into a composite column representing earth history (see fig. 2.3). There is no single area known in which a rock sequence representing all, or even a majority part, of the geologic time scale is present. It was therefore necessary to compile data from many areas and assemble them in vertical sequence to build a complete record. The building of such a scale was done by many workers during the past two centuries. Earth history could be reconstructed only in terms of such a geologic time scale; most important was a means of recognizing the various time-subdivisions of the scale in rocks all over the world. A primary objective of the geologist is the correlation of rock units over great distances. To correlate in geology is to demonstrate equivalency in age; for example, to say that rock unit X and rock unit Y are correlative is to show that they were formed during the same interval of geologic time.

Rocks may be dated or correlated in four different ways. The first dating was done by the **law of superposition,** formulated by Nicolaus Steno in 1669, which states that in a sequence of rock strata which have not been overturned by crustal movement, the oldest beds are at the bottom and the youngest at the top. Naturally the age of a bed so derived would be limited to a statement such as

"Bed B is older than C, but younger than A." If bed B outcropped in a way such that A and B were not visible below and above, then B would not be datable by Steno's law. (See fig. 2.1.)

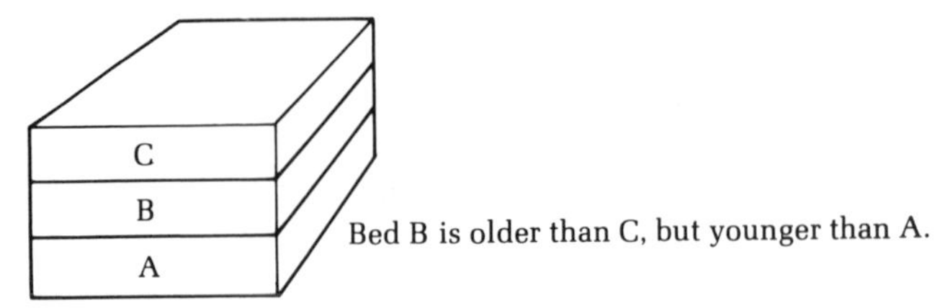

FIGURE 2.1 Steno's Law. The first dating of rocks was done through application of the law of superposition, contributed by Nicolaus Steno in 1669 as a result of his studies of the sedimentary strata of Tuscany in northern Italy.

How Fossils Reveal Age

During the last decade of the eighteenth and first of the nineteenth centuries, William Smith, a largely self-taught surveyor and canal engineer, added the use of fossils to Steno's simple law of superposition and derived a much more comprehensive and widely applicable principle. While mapping in England and Wales, he discovered that each interval of earth history is marked by unique and diagnostic groups of fossil plants and animals which became entombed and preserved within the sedimentary rock beds. His realization that "the same fossil species are found in the same rock layers in the same sequence from place to place" was almost simultaneously discovered by Georges Cuvier and Alexandre Brongniart, working in France. It was this **principle of faunal succession** which led to the development of the geological time scale, for it enabled the observer to date bed B even in the absence of A below and C above, if bed B contained telltale guide or index fossils. This key idea of succession of fossils soon was used by other workers who collected fossil assemblages from areas in which the assemblage could be identified as to its position within a sequence of fossil assemblages. Eventually, sequences of assemblages from many localities in the world were collected and identified, and the fossil species diagnostic of specific geologic time intervals were recognized; if an isolated rock outcrop containing time diagnostic species is found now, it can be placed within the scale of geologic time which evolved from such studies. The changes in the species composition of fossil aggregates representing different intervals of geologic time is due to ongoing organic evolution. Although most species of fossil organisms are too long-lived or narrowly restricted in their geographic distribution, a small number are short-lived and widespread and can serve in dating and correlating rock beds over great distances. Such index fossils are of particular value in attempting to match beds between the continents. (See fig. 2.2.)

We are now able to recognize small subdivisions of geologic time and assign beds to periods, epochs, ages, and even smaller time units if the beds contain

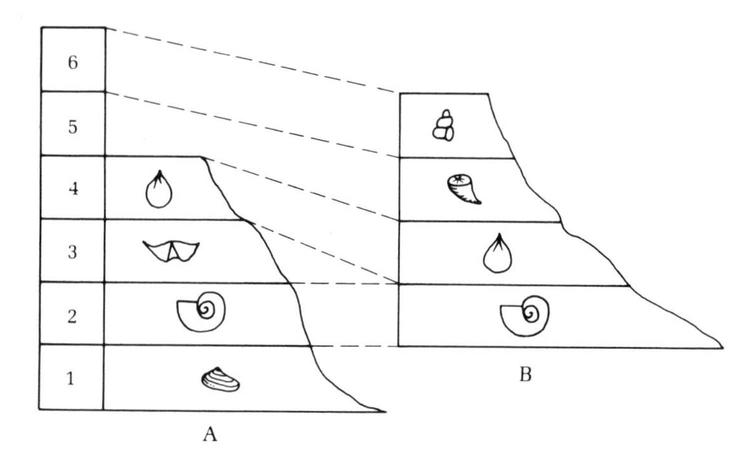

FIGURE 2.2 Correlation by Fossils. Sequences A and B of fossiliferous rock beds are correlated by diagnostic species. Generally, groups of species from each rock bed are used; single species are seldom of value by themselves. Note that bed 3 is present in sequence A but is not present in sequence B; it may never have been deposited at B or perhaps was removed by erosion before the deposition of bed 4.

index fossils. Fossils are our most frequent and important guide in the dating and correlation of rocks.

Although we are greatly dependent upon them in dating, fossils which are both short-lived and widespread geographically, and hence useful in correlation, are not abundant. Usually only a very small percentage of the species in a diverse fossil fauna are guides to small intervals of geologic time. Fossils first appear in abundance from about 600 million years ago at the beginning of the Cambrian, the first period of the Paleozoic era. Before that time at the start of the Cambrian, fossils are very scarce except for a few scattered remains consisting of sponge spicules, worm burrows, some microorganisms, and a few exotic invertebrate types which do not appear to be closely related to the great flourish of life forms which appear in the Cambrian and younger rocks. (See fig. 2.3.) In short, fossils do not permit us to date rocks older than the Cambrian period (about 600 million years old); therefore, we use other means to date Precambrian rocks.

Some dating can be accomplished by **crosscutting relationships,** but such methods are not usually applicable to correlation over long distances. If a molten rock mass invades or cuts a preexisting rock, it bakes the invaded rock at the boundary of the intrusion and marks the baked rock as the older. There are other situations in which rocks can be dated relative to some geologic event such as faulting or erosion, but such dating is quite limited in both precision and geography.

All of the dating methods previously mentioned (Steno's law, law of faunal succession, and crosscutting relationships) yield only relative dates for the rocks in question. Radiometric dating or **radiometry** is the only method which imparts an intrinsic or self-contained time value to the rock and allows the assignment of an age in terms of "absolute" time units such as years.

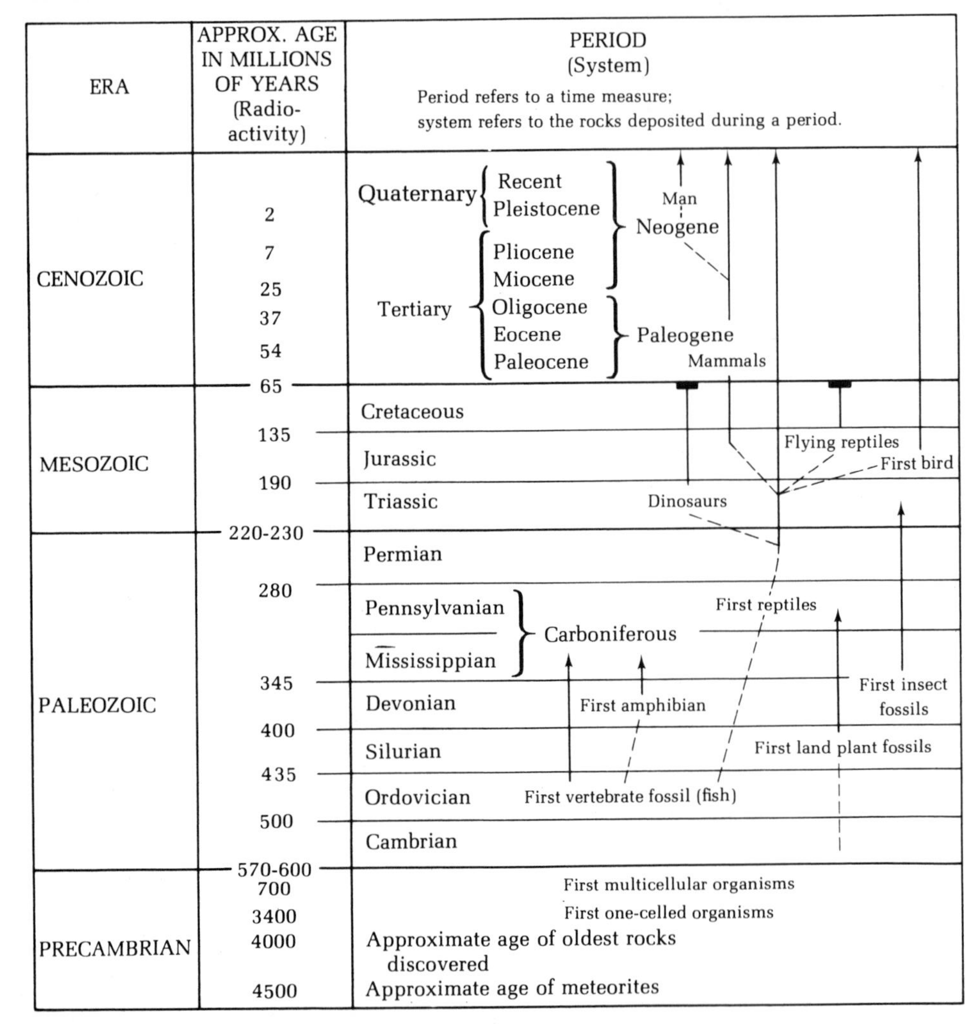

FIGURE 2.3 The Geologic Time Scale and a Simple Schematic View of Some Important Life Groups through Time. The time scale or geologic column is the product of long-term efforts by workers mainly from England, France, Germany, and Italy. As studies of local geology accumulated over wider areas from the mid-eighteenth century onward, the time scale grew more complete. As the sequence of rocks and their contained time-diagnostic fossils became understandable, gaps were filled in and refinements made to earlier subdivisions. Most of the periods in the modern scale were defined in the early part of the nineteenth century. (Reprinted, by permission, from Robert J. Foster, *General Geology*. 2nd ed. Columbus: Charles E. Merrill Publishing Co., 1973. Courtesy of the author.)

Using Atoms to Date Rocks

Radiometric dating is based on the measurement of the ratios between original parent atoms and the daughter atoms which are derived from them through the process of spontaneous decay. In such decay the atom is reduced to a lower energy state. Such radioactive decay takes place in the nucleus of the atom and is independent of conditions in the environment such as pressure, temperature, chemical bonding forces, etc. The decay rates for specific elements are known; therefore, we can expect a certain ratio between parent and daughter atoms after fixed lengths of time. In such a dating method if a mineral grain were examined and found to contain only parent atoms of a given parent-daughter decay pair, the age of the grain would be zero; i.e., no time would have passed yet since the mineral formed in which decay could have taken place. If we know the decay rate for a particular element, we need only to measure the ratio of parent to daughter atoms present. (See fig. 2.4.)

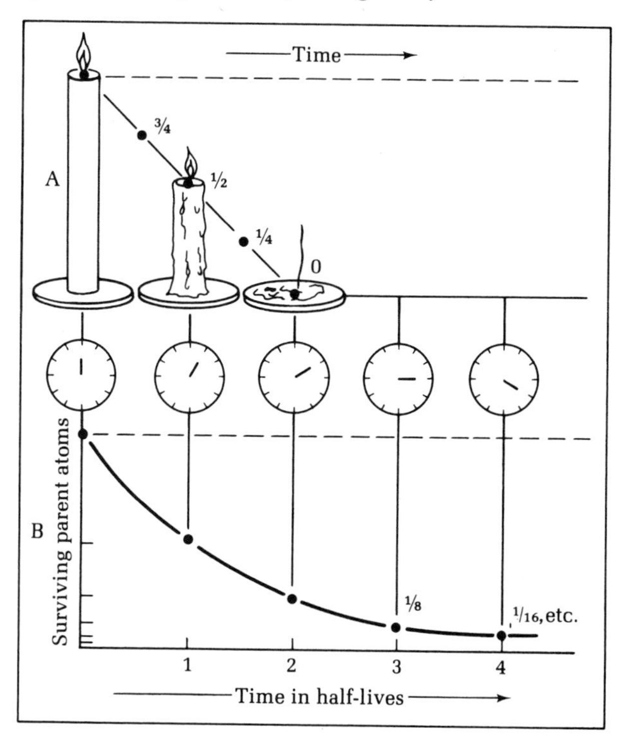

FIGURE 2.4 Half-life. The candle shows a uniform straight-line decrease; in contrast the radioactive decay curve illustrating half-life is quite different. The end of one half-life interval is the start of a new one. (Modified after Eicher, 1968.)

Such a system of age measurement is based on two important assumptions: first, that no parent or daughter atoms have been added to or left the mineral grain in question during the process of decay since the mineral was formed; and second, that no daughter atoms were present in the mineral grain when it formed (all daughter atoms must be the product of decay within the grain). Various sophisticated techniques and methods are required to validate these two assumptions before a sample can yield a reliable radiometric date.

The half-life of an element is defined as the time required for one-half of the original number of atoms of an element to disintegrate into daughter atoms. Only a few of the many radioactive elements occurring in nature have provided most of the radiometric dates which have been assigned to rocks of the crust and facilitated our assignment of time value in years to the subdivisions of the geological time scale (see table 2.1).

TABLE 2.1 The Half-lives of the Important Radiometric Parent-Daughter Pairs. The carbon 14 method, which is widely known because of its applicability in the dating of archaeological materials, is not very useful in geological dating because the half-life of carbon 14 is only 5730 years. Such a half-life limits the method to the dating of materials no older than about 40,000 years.

Parent Atom	Half-life in Millions of Yr	Radiometrically Important Daughter Element	Commonly Dated Rocks and Minerals
Potassium 40	1300	Argon 40	Muscovite Biotite Hornblend Glauconite Sanidine Whole volcanic rock
Rubidium 87	47,000	Strontium 87	Muscovite Biotite Lepidolite Microcline Whole metamorphic rock
Uranium 238	4510	Lead 206	
Uranium 235	713	Lead 207	Zircon Uraninite Pitchblende
Thorium 232	13,900	Lead 208	

Sometimes closely related rock samples contain more than one parent-daughter atom pair, such as potassium-argon, rubidium-strontium, or uranium-lead, and allow independent radiometric checks. Such checks increase our confidence level. As older and older rocks are dated, the error becomes larger. The error in radioactive dating is about 3%; when a

10-million-year-old sample is dated, a 3% error amounts to plus or minus 300,000 years, but when a billion-year-old rock is dated, it may be in error by 30 million years.

It is only during the last few decades that "absolute" dates expressed in years have been assigned to the time scale which originally evolved based only on major physical events and diagnostic fossil species. All of the recent work in radiometric dating is well in agreement with earlier relative dating done by paleontologists, who used fossil faunas and floras as guides to the intervals of earth history. The role of rock dating appears repeatedly as a handmaiden to other areas of geologic investigation which bear on the question of continental drift and plate tectonics.

3

Early Evidence for Continental Drift

Jigsaw Fit of the Continents, Matching Rock Layers, Mountains, and Great Faults

Sir Francis Bacon and Father Francois Placet are often erroneously credited with the earliest remarks (1620 and 1668, respectively) on the jigsaw-puzzle fit between the opposite coasts of the Atlantic. Bacon never compared the coasts facing each other across the Atlantic; instead he merely commented on the west coasts of Africa and South America. Placet believed that an Atlantic island had sunk to produce the Atlantic Ocean and, like Bacon, did not imply horizontal movement of the continents. Alexander von Humboldt was really the first to comment specifically on the jigsaw-puzzle fit between the opposite Atlantic coasts in 1801, but he explained their separation by erosion. It was not until 1858 that Antonio Snider-Pellegrini clearly stated for the first time that the similarity of the opposite coasts of the Atlantic was a proof that America split from the Old World and drifted westward. Snider-Pellegrini's major argument for former connection was based on the very great similarity of fossil plants of Pennsylvanian age (about 300 million years old) found in the coal measures of both North America and Europe.

Wegener opened chapter 1 of his classic work with the remark that "he who examines the opposite coasts of the South Atlantic Ocean must be somewhat struck by the similarity of the shapes of the coastlines of Brazil and Africa. . . ." He noted that "the first notion of the displacement of continents came to me in 1910 when, on studying the map of the world, I was impressed by the congruency of both sides of the Atlantic coasts. . . ." If the present continents are fragments of a once larger landmass, then we should expect to find a number of geographic and geologic features which they originally had in common, including (a) close jigsaw fits between formerly connected coasts; (b) specific rock types of the same age brought into contact on reassembly; (c) continuation of contemporaneous mountain belts from one continent to another; and (d) great lateral faults involving movements of hundreds of miles which end abruptly on

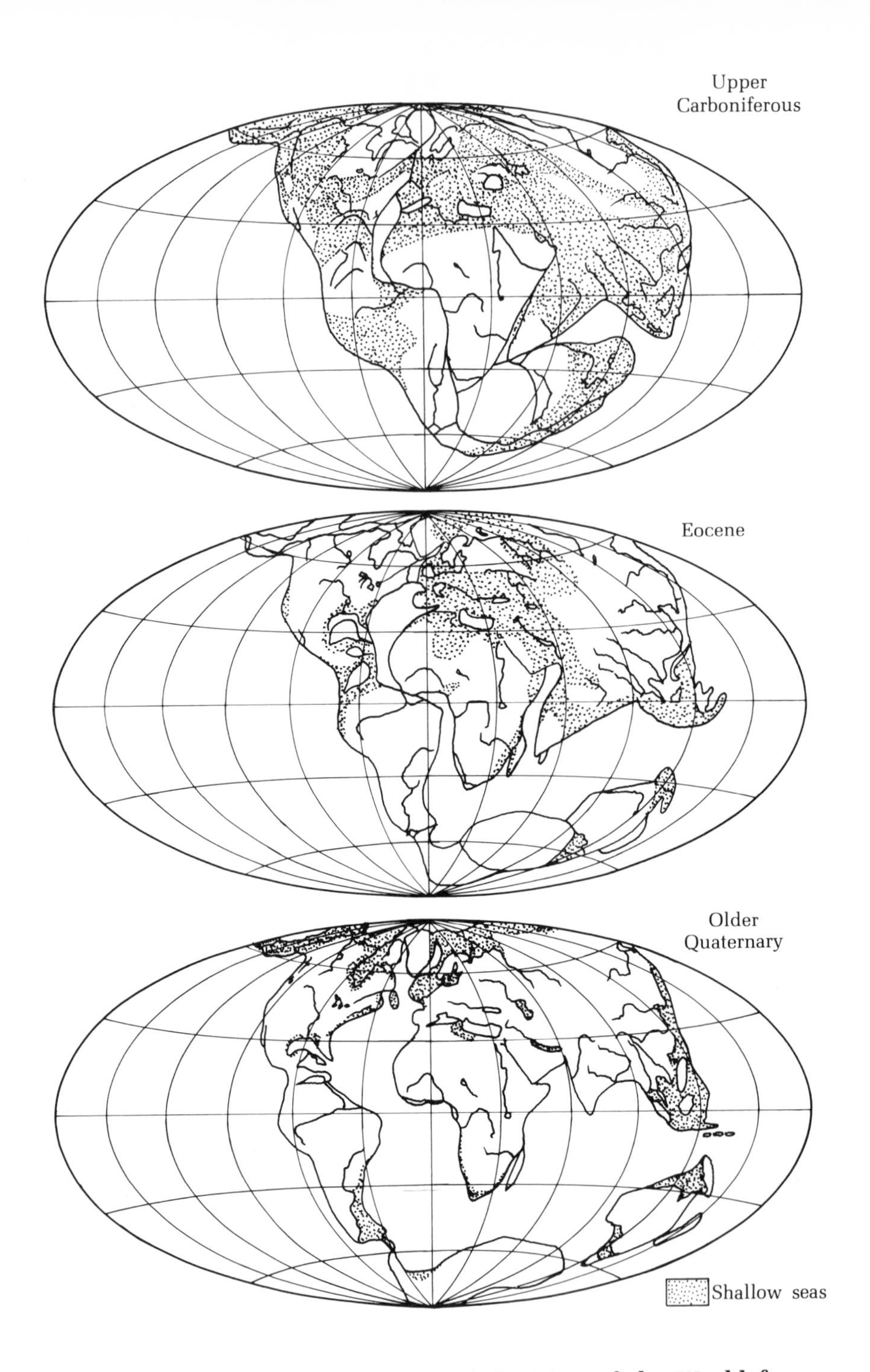

FIGURE 3.1 "Reconstructions of the Map of the World for Three Periods according to the Displacement Theory. Dotted — Shallow seas; present day outlines and rivers only for the purpose of indentification. Latitude and Longitude arbitrary (being that of contemporary Africa)." (Reprinted, by permission, from Alfred Wegener, *The Origin of Continents and Oceans*. London: Associated Book Publishers, Ltd., 1924.)

one continent should continue on another. (The arguments for drift based on the fossil record have been deferred to a separate chapter.) Let us examine these possible lines of evidence in order.

The maps of Wegener's time provided him with a fit between the opposing Atlantic coasts, which moved him to remark that they match "as closely as the lines of a torn drawing would correspond if the pieces were placed in juxtaposition." Wegener's "Reconstructions of the map of the world for three periods according to the Displacement Theory" (fig. 3.1) show shallow seas stippled and rivers for geographic identification. The best recent continental matching has been done between South America and Africa, but it is not to be expected that the original torn edges have retained their shapes perfectly after tens of millions of years of coastal erosion, sedimentation, and crustal wrapping.

Several early attempts were made at fitting the continents together, but they were all simply schematic. In 1958 S. Warren Carey, an Australian, plotted the continents which border the Atlantic on a 2½-foot diameter globe on which the continents could be moved to accomplish the best fits. Carey's fits were not made along the present actual shorelines which have been greatly changed through geologic time, but instead his continental edge was drawn at a depth of about 6000 ft at the base of the continental slope. He noted ". . . if North America, South America, and indeed Antarctica and even Australia are swung as a whole like a gate [around the arc of the Alaskan mountain range from Mt. Logan to Mt. McKinley]. . . through twenty to thirty degrees towards Eurasia and Africa, the motions are such as to close not only the Arctic ocean, but also the Atlantic and Southern oceans." He then continued to draw even more detailed geographic correlations. (See fig. 3.2.)

In 1965 Sir Edward Bullard, an English geophysicist working with J. Everett and A. G. Smith, confirmed Carey's continental reconstructions by a geometrical analysis run on a computer. Bullard's computer approach showed that the

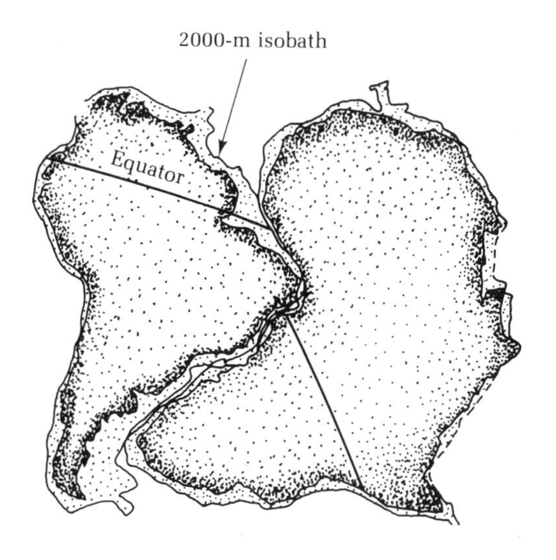

FIGURE 3.2 Africa and South America Joined at the 6000-ft Subsea Contour. (After S. Warren Carey, 1958.)

best fit between the continents could be obtained by drawing coastlines about halfway down the continental slope at a depth of 6000 feet, as Carey had done. (See fig. 3.3.)

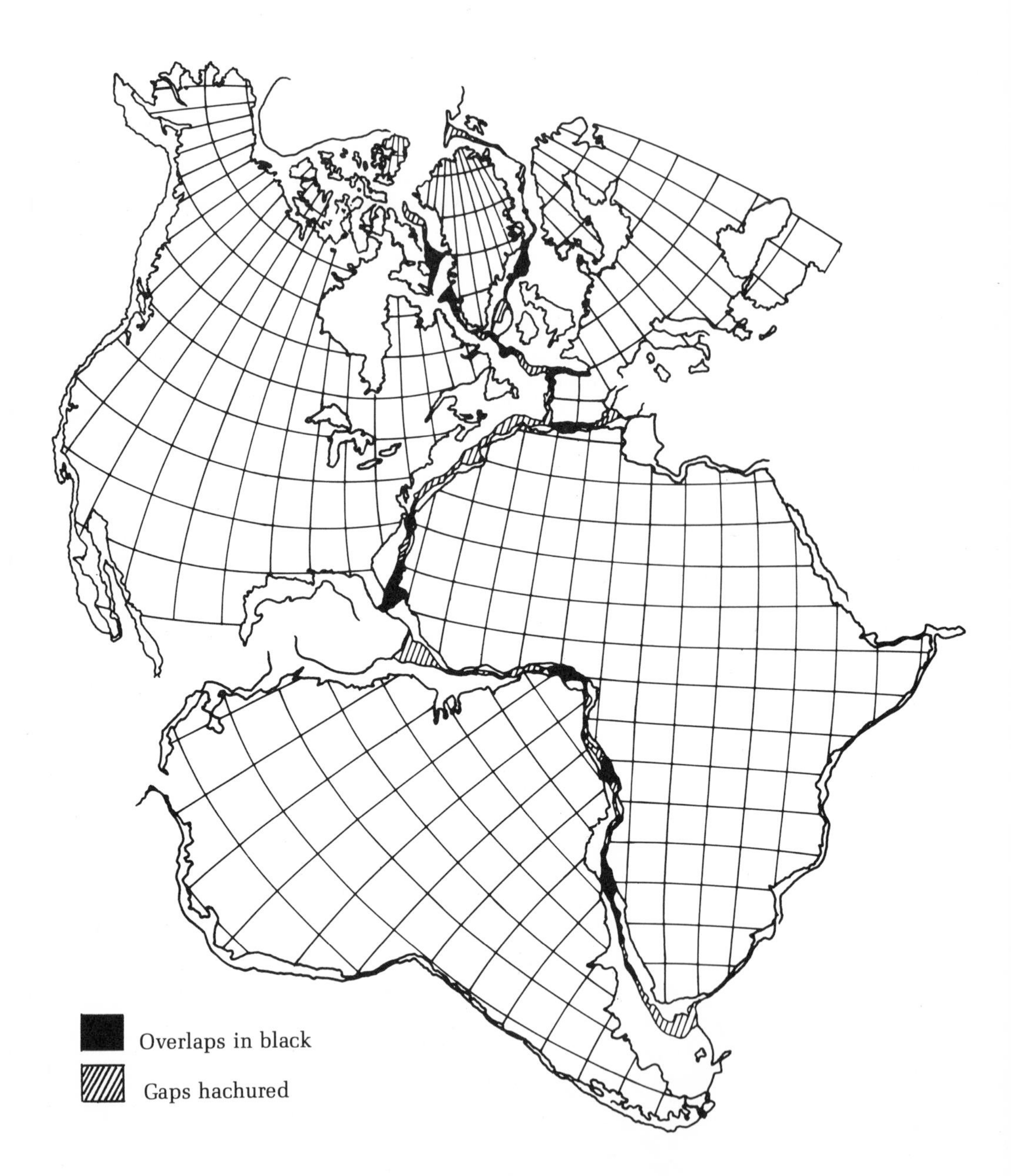

FIGURE 3.3 The Fit of the Atlantic Continents along the 6000 ft-subsea Contour. The figure was done by E. C. Bullard, J. Everett, and A. G. Smith in 1965. The mismatch (overlap or gap) over most of the reconstructed boundary averages less than one degree. (Modified after Bullard et al. *Philosophical Transactions of the Royal Society of London* 258, 1965.)

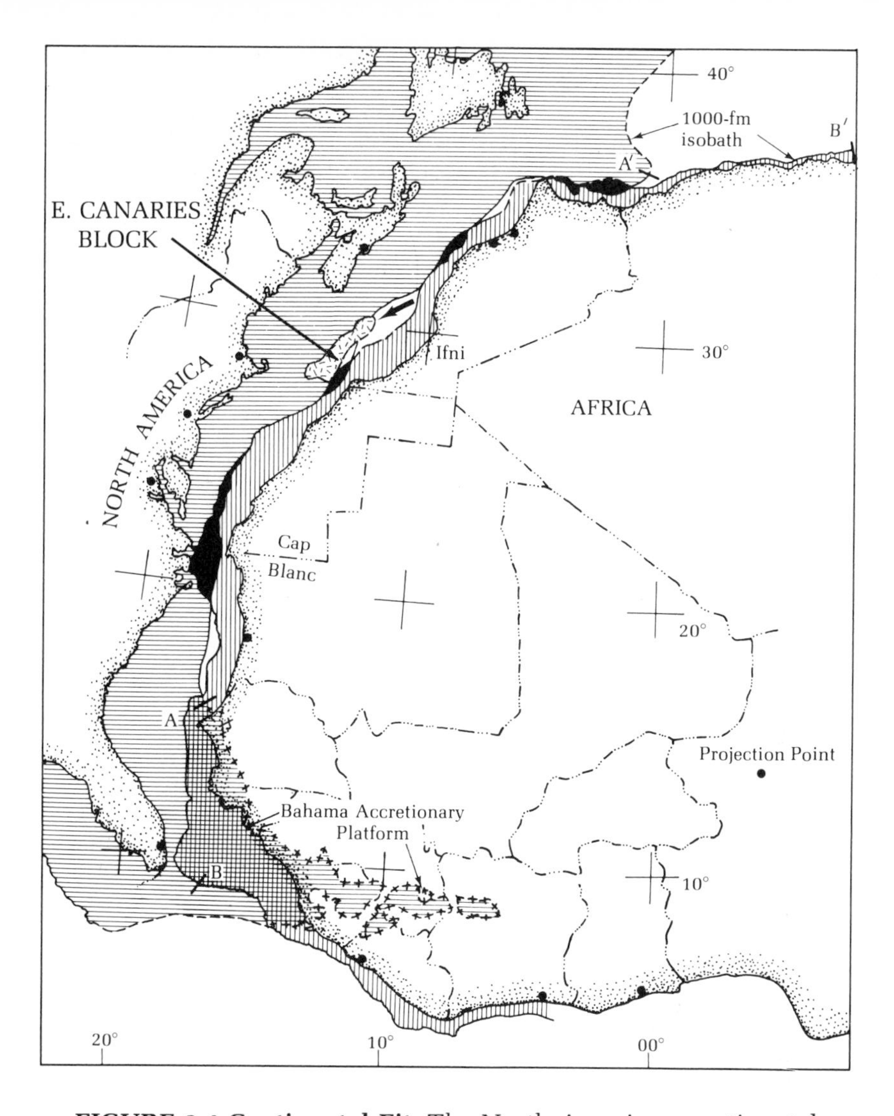

FIGURE 3.4 Continental Fit. The North American continental edge at 6000-ft depth (1000-fm isobath) between A and A[1] was fitted against the African continental boundary at the same depth between B and B[1]. The North American continental shelf is shown by horizontally ruled lines and the African shelf vertically ruled lines. White areas are gaps in the fit; black represents overlaps. The great area of the Bahamas platform (shown hachured) came into being after continental separation through the deposition of sediments followed by the growth of coral reefs. The indentation in the African coast at Ifni marks the largest gap in the fit; it may have been formed when a crustal segment broke off and drifted 115 miles southwest to form the eastern Canary Islands. (From R. S. Dietz and W. P. Sproll, Proceedings ICSU/SCOR Symposium, Geol. of the East Atlantic Margin. Cambridge, Institute of Geol. Sci. Report No. 70/13, 1970. Crown Copyright Geological Survey diagram. Reproduced by permission of the Controller of Her Britannic Majesty's Stationery Office.)

Both Carey and Bullard show a very close fit between Africa and South America with the misfit amounting to only a few percent. It is thought that the perfection of the fit may be due to the tough character of the rock which was split when the continents separated. The break occurred through an ancient (Precambrian) continental nucleus area in which the rocks had been toughened and made more resistant to erosion by metamorphism, a process of profound change due to the long-term application of heat and pressure. To the north, there are complex mountain belts more than 200 million years old which lie along the line of continental separation in North America, England, and eastern Europe and which thus do not allow as close a fit between those areas.

Using a computer, Walter P. Sproll has recently fit northwestern Africa against North America. He, too, assumed that the actual continental edge lies at about 6000 feet down the continental slope (see fig. 3.4). Although not nearly so perfect a fit as the remarkable one between Africa and South America, Sproll's reconstruction certainly appears plausible and not likely the result of random matching.

Recently several other workers have turned to computer reconstructions of the continents. The most comprehensive reassembly has been done by Robert Dietz and John Holden, who drew upon the work of Sir Edward Bullard, J. E.

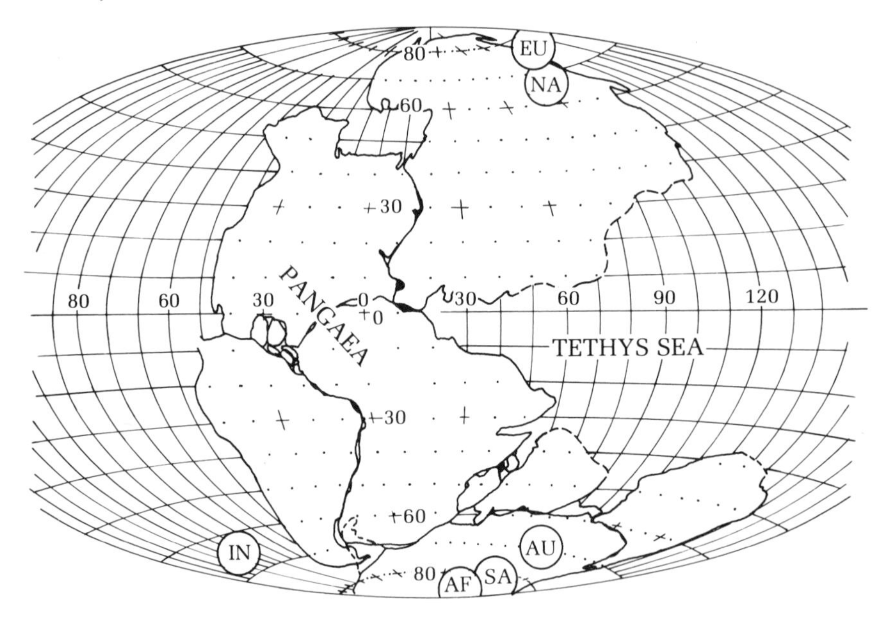

FIGURE 3.5 Pangaea. This great single landmass existed on earth at the end of the Permian period 225 million yr ago. The relative positions of the continents, with the exception of India, have been worked out by computer to get the best fit. The continental edges have been drawn about halfway down the continental slope at a depth of 6000 ft since that is thought to be half the height of the cliffs formed when the continents first split apart. (Reprinted, by permission, from Robert S. Dietz and John C. Holden, *Journal of Geophysical Research* 75, p. 4942, 1970, copyright by American Geophysical Union.)

Everett, and A. G. Smith of Cambridge University for uniting the Atlantic Ocean continents. They credit Walter Sproll for his computer solution in closing the Indian Ocean which involved good fits between Australia, Antarctica, and Africa. These three continents comprise the ancient landmass of Gondwana, which we refer to in later chapters. India is also thought to have been a part of Gondwana, but its position of attachment has not yet been resolved. The new methods of reconstruction employed by Dietz and Holden are discussed in chapter 7, and their reconstruction is shown in fig. 3.5.

Matching Rock Formations

In his chapter on geological arguments for drift, Wegener pointedly remarked, "A very sharp control on our supposition . . . is exercised by a comparison of the geological structures on both sides [of the Atlantic]; for it is to be expected that folds and other structures, formed before the separation will be continuous from one side to the other, and their terminations on both sides of the ocean must lie exactly in such a position that they appear as immediate continuations in the reconstruction." He cited H. Keidel's work of 1916 in which it was shown that the sequence of rock beds in the Sierras of the Province of Buenos Aires are closely comparable to the succession of strata in the Cape Mountains of South America. These strata are involved in a long, ancient, folded mountain belt which cuts the southern point of Africa and extends into South America south of Buenos Aires, turning northward to connect with the Andes. Wegener considered the former connection of the Cape Mountains of southern Africa with the Sierras of Buenos Aires only one of several such proofs along the coasts of the Atlantic.

Wegener also later drew upon the 1921 work of H. A. Brouwer, who compared the igneous rocks of Brazil with those of southern and southwestern Africa and found five different major rock bodies which were correlative and bestowed "a similar topographical character on the landscape in both continents." Brouwer also favorably compared the sedimentary rocks, noting that the South African Karoo matched the Santa Catharina rocks of Brazil.

Alexander Du Toit, the eminent South African geologist, has played a major role in the drift controversy. He assembled much evidence showing intercontinental correlations in a comprehensive book titled **Our Wandering Continents** of 1937. Du Toit's comparison of fossil-bearing rocks between South America and Africa show that the oldest such rocks are late Silurian or Devonian age (about 400 million years old) and contain fossils which differ from those of the Northern Hemisphere but which are similar to those of the Falkland Islands. Both areas also show an even more significant succession of late Paleozoic- to Jurassic-age beds called the Gondwana, which contain unusual, very mixed conglomerates deposited by massive continental glaciers; such rocks are called **tillites.** These unusual rocks of about the same age are also found in India, Australia, and Antarctica (see fig. 3.6).

The Gondwana tillites of all of these southern continents contain groups of fossil plants or floras dominated by the genera **Glossopteris** and **Gangamopteris** which aid in their dating and indicate that the glaciation was probably contemporaneous on the southern continents. In 1962 G. A. Doumani

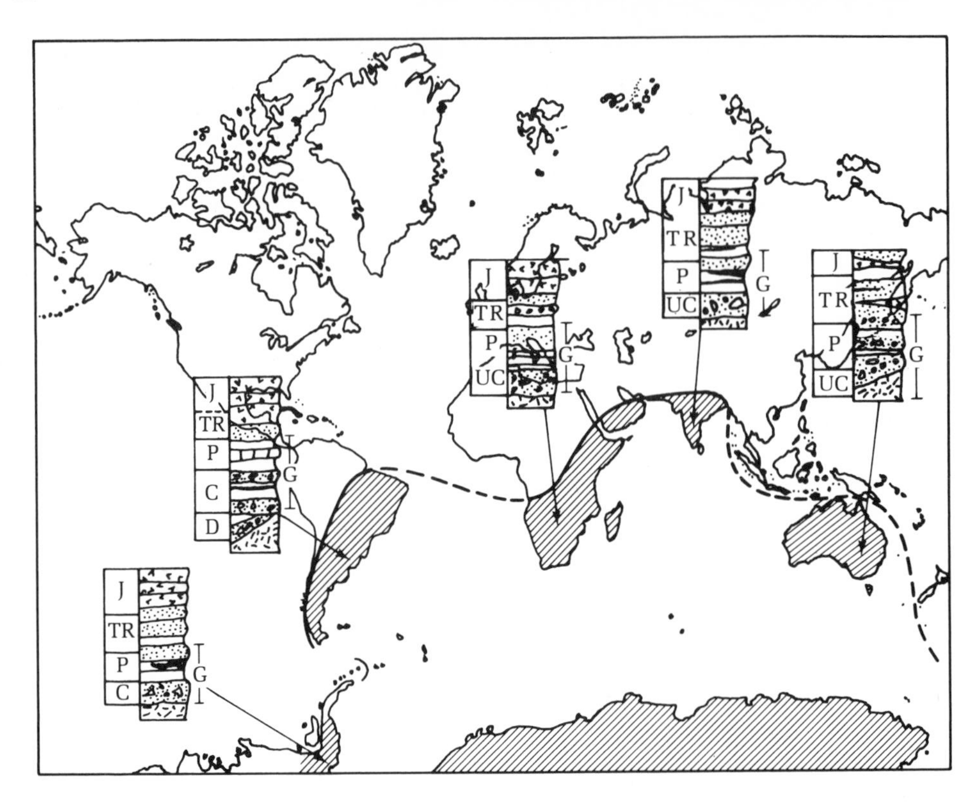

FIGURE 3.6 The Similarities of Rock Layer Sequences of the Same Age from the Gondwana Continents. The records from Africa, Antarctica, Australia, India, and South America of Carboniferous-Jurassic age are almost identical. Such evidence from widely separated continents suggests that they were all formerly joined to form Gondwana, the great southern landmass. (Reprinted, by permission, from Leigh W. Mintz, *Historical Geology: The Science of a Dynamic Earth*. Columbus: Charles E. Merrill Publishing Co., 1972. Courtesy of the author.)

and W. E. Long reported that they had uncovered the same plant fossils in tillites of the Antarctic, lending further support to Du Toit's Gondwana argument. As the glaciers moved across exposed bedrock, the cobbles and boulders frozen into the glaciers produced striations or scratches and thus left clear indications of direction of movement. (See fig. 3.7.) In addition, the rock debris picked up and often carried great distances can be traced to its place of origin and provide further information about the glacier's travels. This kind of evidence indicated that glaciers moved across Australia from the south and penetrated the Falkland islands from the north. A glacier centered near 5° south in Africa left its northern deposits on the equator and its southern deposits near the Cape of Good Hope. Another great ice mass was located north of Bombay in India and moved hundreds of miles to the northwest and southeast. The glaciers passed northwestward onto South America. Brazilian tillites contain rock types found in southwest Africa, but they are not found elsewhere on the South American continent! There is much additional evidence from many localities on the southern continents which demands that we unite the southern continents or alternatively explain how glaciers were miraculously born in tropical temperate oceans and there picked up and carried continental-type igneous rocks onto the southern continents.

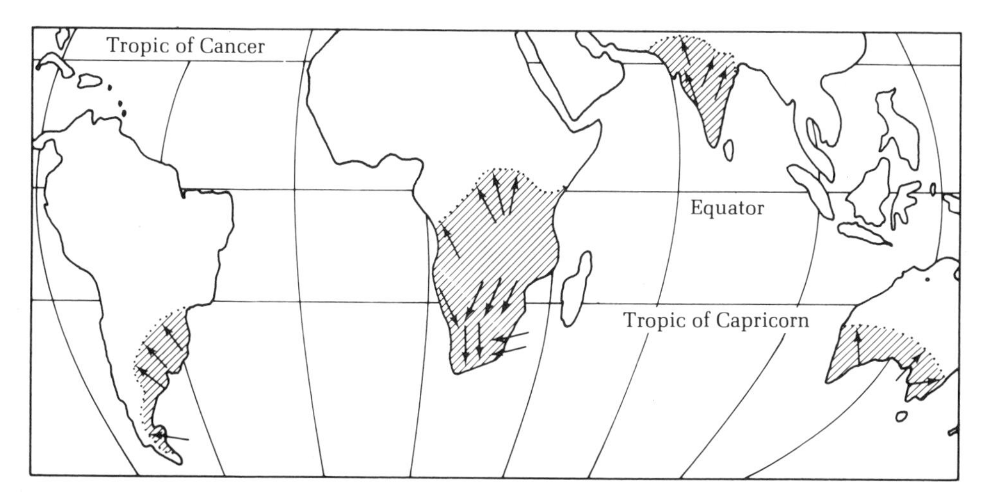

FIGURE 3.7 The Directions of Glacier Movement and Distribution of Late Paleozoic (about 250 Million Yr Old) Glacial Deposits. (From Mintz, *Historical Geology.*)

A recent re-examination of the geological evidence was done in 1961 by H. Martin in an attempt to discern if the more detailed information which has accumulated since Wegener and Brouwer's time has increased or decreased the number of known similarities between Africa and South America. He sounded the conclusive note that "there is not the slightest doubt that from the Silurian to the Cretaceous every correction of the stratigraphy [study of stratified rock beds] and lithologic [rock] columns has increased the similarities. The stratigraphic and lithologic columns for this period of some 200 million years have become almost identical. I do not think that, for a comparable length of time, a similar likeness between parts of any two continents can be found." (See fig. 3.6 and 3.8.)

About 500 to 800 million years ago, a major orogeny, or mountain-building movement, occurred which greatly deformed the crust in the area along which South America and Africa split. The great deformation gave a "grain" to the region, trending roughly parallel to the present coasts, which probably influenced the fracturing pattern.

Recent advances in radiometric dating of rocks have allowed the intercontinental matching of both rocks and structures which were previously in question. Radiometric dating (discussed in chapter 2) is based on the measurement of the ratios between original parent atoms and the daughter atoms which are derived from them through atomic decay which goes on at a constant rate. If we know the decay rate for a particular element, then we simply measure the ratio of parent to daughter atoms present. Such a system of measurement is quite complex and has many limitations in application, but many rocks now lend themselves to such dating.

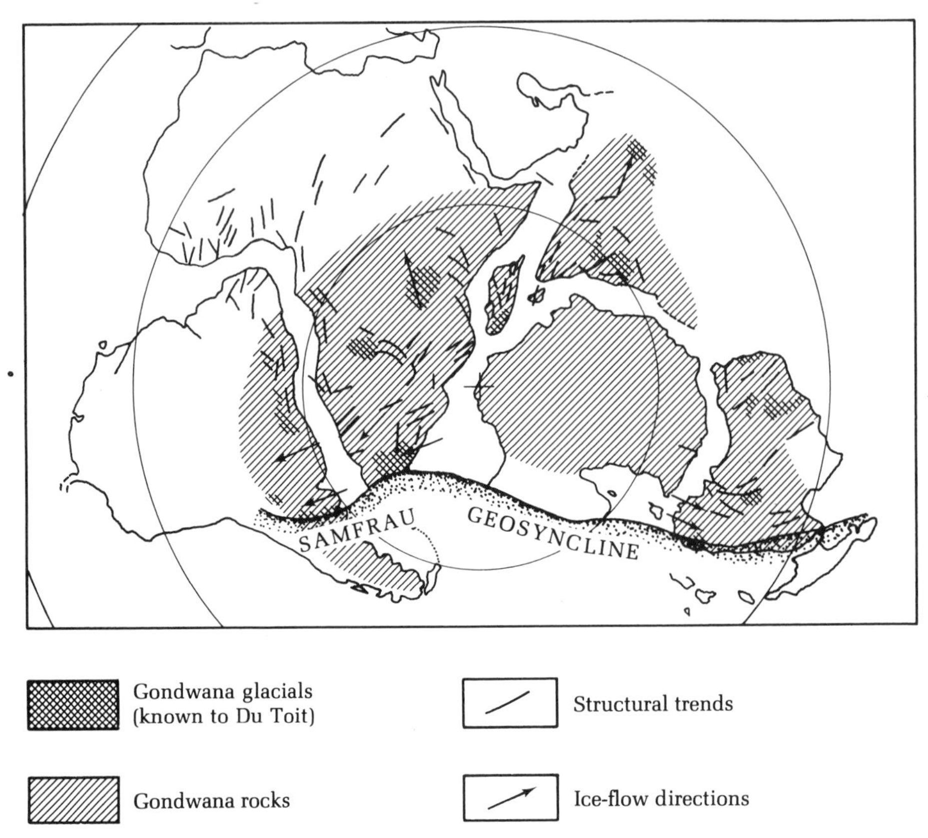

FIGURE 3.8 Gondwanaland as It Appeared about 250 Million Years Ago at the End of the Paleozoic Era. The extent of glacial deposits, ice-flow directions, and alignment of mountain belts are shown. (After Du Toit, 1937, and from Mintz, *Historical Geology.*)

In Africa there is a clearly defined boundary between two large areas underlain by rocks of very different character and age (see fig. 3.9). The boundary trends roughly northwest-southeast, passing near Accra in Ghana, and separates 2000-million-year-old rocks of the Eburean-age province in the north from 550-million-year-old rocks of the Pan African-age province to the south. P. M. Hurley of the Massachusetts Institute of Technology dated these rocks in Africa and then sought a similar boundary between comparable rock provinces in South America; of course, his search in South America was aided by the jigsaw fits of the two continents. He found what he was seeking near São Luis in Brazil, the exact location indicated by the jigsaw fit. Hurley dated a great number of rock samples along the coastlines of each continent, and he found

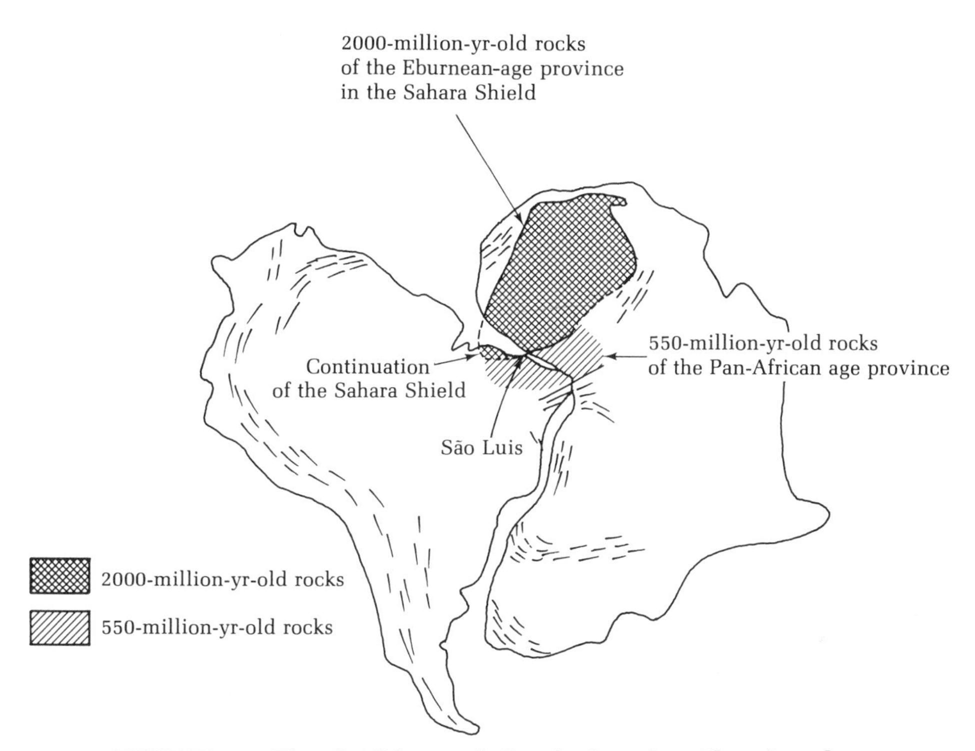

FIGURE 3.9 Fit of Africa and South America Showing the Continuation of Rock Masses of Identical Age and Belts of Deformed Rock. (After Hurley, 1968.)

that the rocks matched in age; additionally, the structural trends or grain in the African rocks have their continuation in South America when the continents are reassembled.

Matching Folded Mountain Ranges

The trends of major folded mountain ranges provide another test of proposed continental reconstructions. During certain intervals of earth history, portions of the crust have been compressed to form mountain ranges hundreds or thousands of miles in length. The rocks forming them were originally sediments in which animals were entombed, and their fossils now permit their dating. Additionally, the composition and size of the sedimentary particles and structures within the sedimentary beds often tell the direction from which the sediment was carried by streams. On that basis, North American geologists long believed that the sediment in the rocks of the Appalachian Mountains was derived from a landmass which lay to the east in the present site of the Atlantic Ocean. The sediments in the rocks of the Caledonian Mountains of Britain were derived from the north and west, indicating that there, too, land once lay in the site of the present Atlantic Ocean. Geologists formerly believed that such offshore landmasses had sunk after they supplied sediments to build the Appalachians and Caledonians. We now know that when the puzzle pieces are fitted together, the Appalachians and Caledonian mountains form a continuous belt. Apparently there were several different episodes of crustal disturbance in different places along the belt, but the unity of the belt is clear. Continuities in the trends of folded mountain belts are to be expected if the continents are restored to their predrift positions.

There are many other such places on earth at which the indicators for sediment source point to what must be a missing landmass. Martin's study of the Devonian sedimentary rocks of the Paraná Basin of Argentina and the direction of glacier movement near Capetown, South Africa, both indicate that raised landmasses must have occupied the present site of the South Atlantic.

Africa and South America were both flooded by basaltic lava outpourings at the start of the Jurassic (about 180 million years ago). They accumulated to a thickness of almost 5000 feet, covering about one-half million square miles. Although much of these volcanic basalts have been eroded in southern Africa, evidence shows that they originally covered an equally large area there. It is thought that such massive volcanic activity signaled the splitting of the protocontinent of Gondwanaland. Evidence from ancient magnetism (discussed in chapter 6) preserved in the basalt rocks of both continents indicates that they began to separate while these basalts were being erupted.

Figure 3.10 shows the pattern of three different groups of folded mountains segregated by age. The belts were folded into existence before separation of the continents. The Late- and Mid-Paleozoic ranges line up nicely across the North Atlantic when the jigsaw pieces are fitted together, but the curved trends of the younger mountain belts do not show such continuations, as is expected, because the continents had separated by then.

When North America and Europe are united according to all of the recent evidence, including paleomagnetism, the margins of the ancient Caledonian Mountains of Greenland and the Northwest Highlands of Scotland very nicely join up with the corresponding margin of the Appalachians in North America. Additionally the late Paleozoic Hercynean Mountain belt and early Paleozoic Caledonian belt which converge westward across Europe and almost meet in southwest Ireland continue nicely into North America where the belts eventually cross.

Matching Cracks in the Crust

In 1962 J. Tuzo Wilson suggested that the Cabot fault, which is really a narrow belt of lateral faults which have been mapped from Newfoundland southwest across the Bay of Fundy and into Massachusetts, is a continuation of the Great Glen fault of Scotland. There the Great Glen fault zone is broad, containing crushed rocks which have been eroded easily to form a string of lakes marking its northeasterly trend across Scotland. It is a left lateral fault along which the rocks can be matched on the two sides, showing an offset of 65 miles. Most of the movement occurred during the late Devonian and Mississippian periods (about 350 million years ago), but it is still actively producing minor earth-

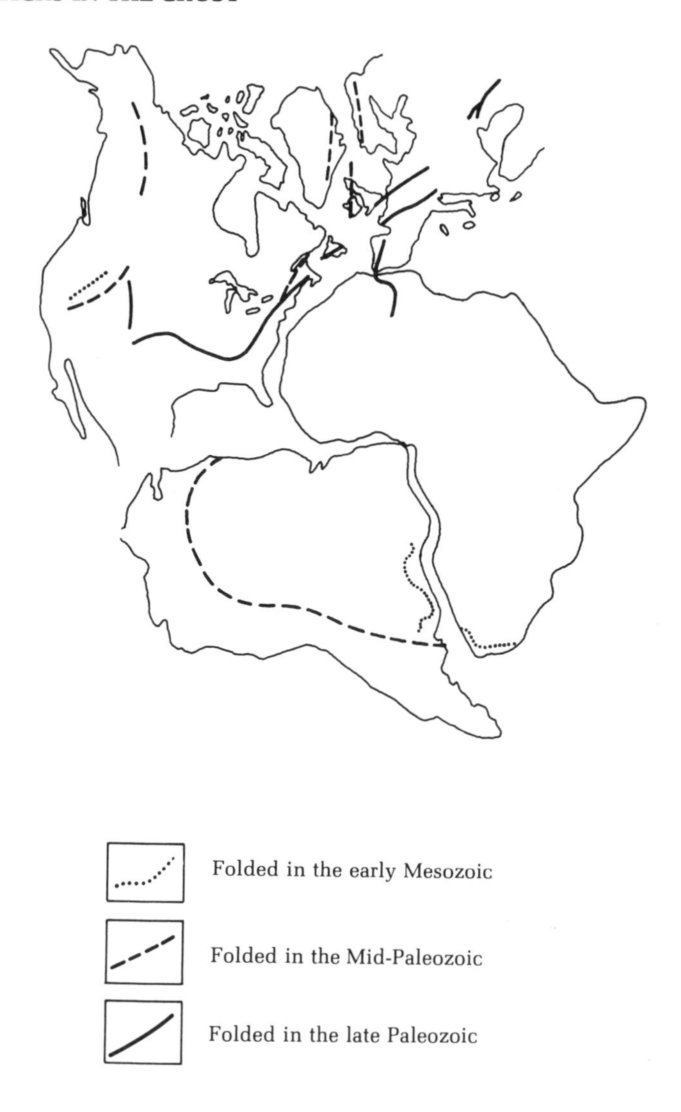

FIGURE 3.10 Folded Mountain Belts of Three Different Ages on the Restored Atlantic Continents. (Reprinted, by permission, from *Principles of Geology*, Third Edition, by James Gilluly, Aaron C. Waters, and A. O. Woodford. W. H. Freeman and Company. Copyright © 1968.)

quakes. The Cabot and Great Glen fault zones are similar and also of the same age; they become well aligned when the continents are reassembled (see fig. 3.11) and provide strong evidence for drift.

The early evidence for drift based on old classical geology provided strong arguments in its own right; one may rightfully wonder why the concept was so strongly opposed for so long in the face of it.

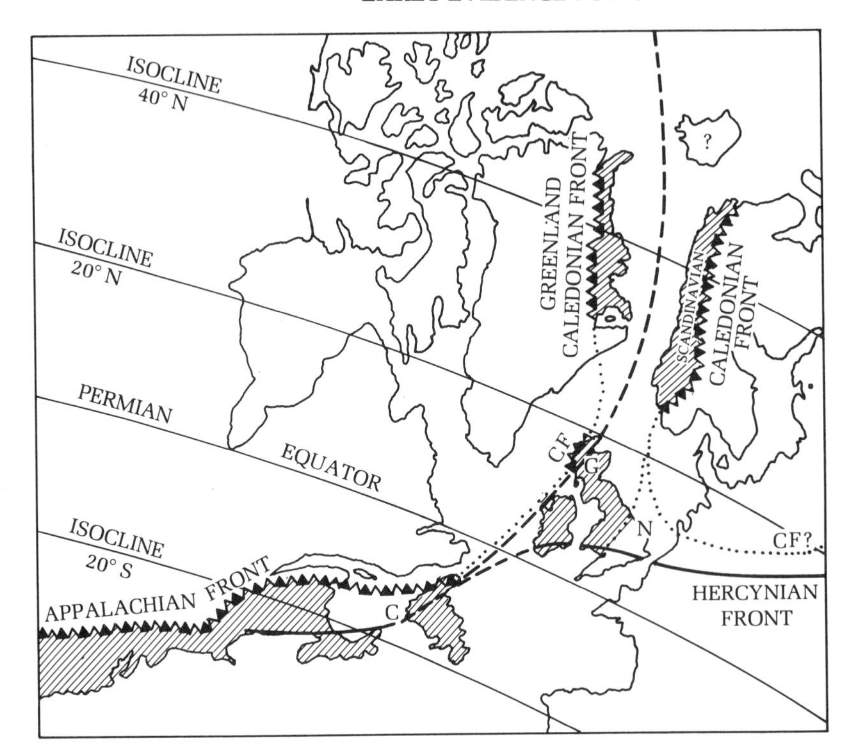

FIGURE 3.11 Reassembly of the Opposing Lands of the North Atlantic Based Mainly on the Evidence of Ancient Magnetism. The *isoclines* are lines of equal magnetic inclination (see chapter 6) and indicate that about 270 million years ago during the Permian, Britain was within 20° of the position of the equator. Europe and Britain therefore must have been thousands of miles south of their present positions, suggesting that clockwise rotation took place during their northward drift. The Appalachian and Caledonian mountain belts are stipled. C F, the Caledonian front of the Northwest Highlands of Scotland; G, the Great Glen fault of Scotland; C, the Cabot fault of Newfoundland and Nova Scotia and its suggested extension along the coast of New England; N, a possible middle area (including part of the North Sea, East Anglia, and the Netherlands) if the Scandinavian Caledonian front should continue through Belgium and Poland. (Reprinted, by permission, from Arthur Holmes, *Principles of Physical Geology*. 1965 Edition. Copyright © 1965, The Ronald Press Company, New York.)

Fossil Life and Climates of the Past

Fossil Life and Environments

The fossil animals and plants found in the sedimentary rocks of the earth's crust tell us much about the conditions under which they lived and died. Each kind of animal and plant is uniquely and finely adapted to the home in which it lives; the science of ecology, which treats the relationship between an organism and its environment, is named from the word **ecos,** which means **home** in Greek. Some species are widely tolerant of a great range of variation in the conditions of their home; others are sensitive to even small changes in such factors as temperature, salinity, moisture, etc., and are forced to migrate in order to escape a changing environment or perish when unable to do so. Some species are widely distributed geographically, adapted to live over a wide range of different environments; others are quite limited in geographic distribution and found restricted to only specific kinds of environments.

The interpretation of ancient environments based on fossil faunas and floras is called **paleoecology.** The central principle in paleoecology, as in other branches of earth science, is **uniformitarianism,** the idea that all of the forces and processes operating on earth today, including those applicable to the existence of living things, have operated in much the same way throughout geological time. The rivers have always flowed to the sea in response to gravity, winds have always resulted from movement of air from high- to low-pressure areas, and fossil animals and plants, which closely resemble living forms, are assumed to have lived under conditions similar to those of their descendants.

Some kinds of organisms are not useful in reconstructing ancient environments and climates because they are widely tolerant of a great range of living conditions; others which have very narrow and specific limits to the conditions under which they can live and reproduce often allow us to make detailed interpretations of ancient landscapes and seascapes.

The clarity with which we visualize an ancient environment depends greatly on the number of different kinds of fossil organisms present. One species tells us something; two, something more; and a great number, a great deal more. In addition to the fossils themselves, the nature and distribution of the rock in which they occur also often reveal much about the past. Such information based on the rock record will be discussed later in this chapter

Land plants are probably the most reliable and broadly distributed life forms of use in interpreting past climates; they are more useful than the vertebrate and invertebrate animals, since they are generally more dependent on and sensitive to their environments.

Fossil Evidence Concerning Drift

Plant evidence was stressed by Wegener in his great work of 1915 which truly opened the revolution in thinking about drift, but much earlier, in 1858, Antonio Snider-Pellegrini argued for reassembly of the continents based on the close similarity between large numbers of fossil plants in the Pennsylvanian period (325–270 million years ago) of Europe and North America. Wegener's main plant evidence came from an exceptional grouping of rocks ranging in age from late Carboniferous to early Triassic (about 200–300 million years ago) called the **Gondwana "System."** The type representative of the Gondwana is located in the Gondwana Province of peninsular India, where glacial tillites, plant fossil beds, and other terrestrially deposited rocks are found. Closely similar sequences of rocks of the same age are found in central and southern Africa, eastern South America, southeastern Australia, Antarctica, and some islands in between. When all of these places are assembled together in a single landmass, they are called **Gondwanaland.**

The coal measures of the Gondwana and its equivalents on the other continents contain a unique, very poor, and apparently stunted fern flora associated with clear geologic evidence of glaciation. The flora is marked by a small number of species and is very different from the probably contemporaneous coal-swamp flora of the northern landmass. The Gondwana flora is typified by the distinctive genera **Glossopteris,** which has a strong midrib, and **Gangamopteris,** which lacks such a rib (see fig. 4.1). In addition, this flora in the lower part of the Gondwana contains many distinctive and easily identifiable genera. Based on fossil leaves, 20 of 27 species in the Glossopteris flora of Antarctica are found in India. The question of possible intercontinental transport of seeds has never been definitely decided, but it appears unlikely that the seeds of such plants as **Glossopteris,** which are several millimeters across, could have been carried such great distances by the wind.

The first description of the flora from India and Australia was done in 1828, and equivalent forms from South America were reported in 1858. Floras from other localities followed later, culminating in Edward Suess' coining of the term **Gondwanaland.** Suess believed that the Tethys Sea, of which the Mediterranean is a small relic, separated Gondwanaland, the great southern protocontinent, from Laurasia, the northern landmass.

Wegener noted that in Spitsbergen, which is under ice in a polar climate, one can find fossil floras of lower Tertiary age (about 65–40 million years ago), containing a richer variety of plants than is now found in central Europe, for example, firs, pines, yews, limes, beeches, poplars, elms, oaks, ivy, hazel, ash, and many others, including warmth-loving plants such as water lilies, swamp cypress, magnolia, and chestnut. He concluded that a climate such as that of France today must have prevailed in Spitsbergen with the average annual temperature about 20°C higher than at present. He dug in still older rocks in Spitsbergen, finding tropical sagopalms, ginkgo, and tree ferns in the Jurassic and lower Cretaceous (100–180 million years ago). In the Mississippian-age rocks (about 325 million years old) he found a flora identical to that of the great coal deposits of the upper Carboniferous of Europe.

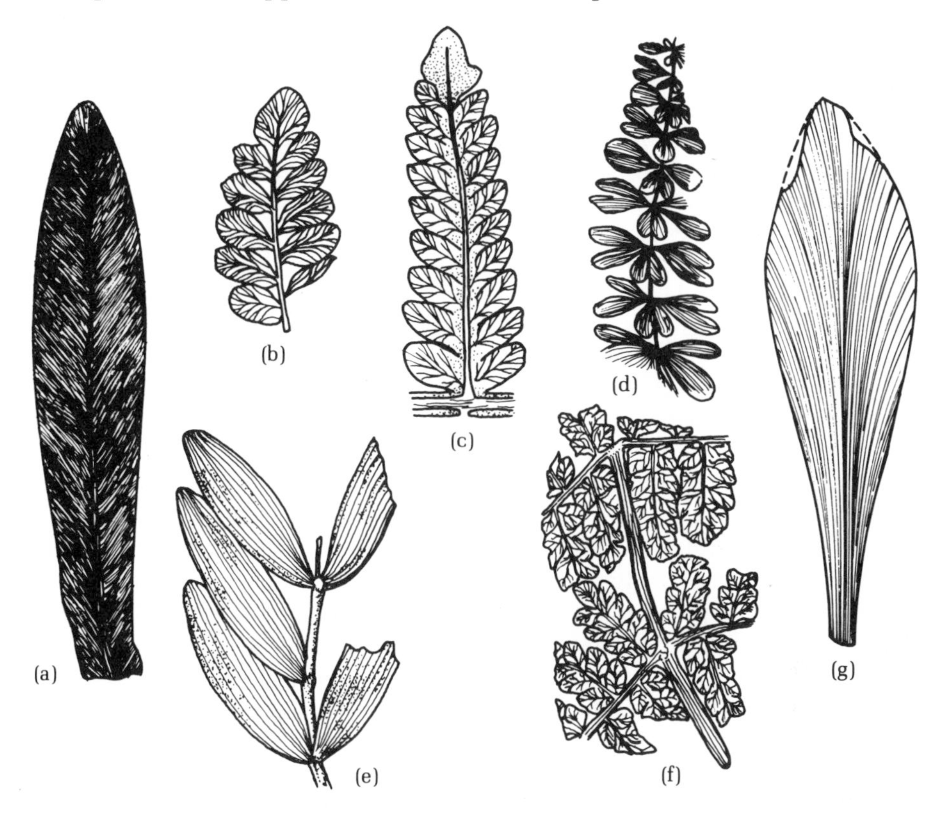

FIGURE 4.1 Representatives of the Gondwana Flora (Glossopteris Flora). A. *Glossopteris* decipiens. **B.** *Merianopteris* major. **C.** *Sphenopteris* polymorpha. **D.** *Sphenophyllum* speciosum. **E.** *Schizoneura* gondwanensis. **F.** *Sphenopteris* polymorpha. **G.** *Gangamopteris* cyclopteroides. Some principal species of the flora are shown in A–G. The flora (fossil plant assemblage) first appears just above the glacial tillites of the late Paleozoic Gondwana rocks and has been found in South America, South Africa, India, Australia, Antarctica, the Falkland Islands, and Madagascar. Its distribution provides evidence for Paleozoic reassembly of the southern continents. [From E. A. N. Arber, British Museum Catalogue of the Fossil Plants of the *Glossopteris* flora in the Department of Geology, 1905, British Museum, London. By permission of the Trustees of the British Museum (Natural History).]

Although some of Wegener's paleobotanical conclusions regarding climate and temperature minima for the past may be somewhat overstated for the evidence which he cited, in the main they point to radically different climates from those which presently prevail at that latitude.

The third Byrd expedition to Antarctica collected fossil plants and coal from Jurassic rocks (about 150 million years old) on Mount Weaver at an elevation of 10,000 feet, only 210 miles from the South Pole — surprisingly, the plants belong to genera found in Jurassic sediments of temperate and Arctic regions of the Northern Hemisphere. Such a find demands an answer to two very difficult questions which seem best answered by continental drift and/or polar wandering. How did coal-forming swamps develop close to the South Pole, and how did a rich and varied Jurassic flora develop under the conditions of the long polar night? Such a group of plants surely would have required light not afforded by the present latitude of the fossil deposit!

Some fossil plant evidence does not favor drift; the American paleobotanist, Axelrod, has long argued that the totality of fossil plant distribution requires that the earth had a climatic zonation essentially like that about our present equator, since land plants first appeared in the Silurian period (a little more than 400 million years ago). Paleobotanists, as a group, have been far from a consensus regarding their evidence and how it speaks on the question of drift, partly because too often drift produced conditions similar to those created by the opening of the Atlantic, in which the east-west movement caused small changes in climate which did not stimulate great evolutionary change and thus had a minor effect on the two separated floras. In contrast, India's Cenozoic, northward drift through 50° of latitude produced severe climatic change which almost wiped out a diverse Mesozoic, gymnosperm (mostly evergreen) flora.

Hallam (1972) has nicely summarized patterns of fossil distribution in the following four categories and related them to continental drift: (1) **Convergence** refers to an increase in similarity between faunas of formerly separated areas such as North and South America which were disconnected for most of the 65-million-year Cenozoic era. During this time each continent in isolation evolved a very distinctive land fauna. About 2 million years ago at the end of the Pliocene, the Panamanian isthmus rose, and a major intercontinental exchange of land mammals produced 22 families in common between the continents. During the earlier time of continental separation about 29 families of mammals lived south of the isthmus and about 27 different families lived in North America. (2) **Divergence** is decreasing similarity between faunas such as occurred among the bottom-dwelling invertebrates of the ocean area which was severed by the rise of the Isthmus of Panama. A formerly homogeneous population was divided at the end of the Pliocene, and evolution has since produced significant differences between the resulting two faunas on the opposite sides. (3) **Complementarity** involves a complementary reaction by both marine and terrestrial faunas in response to the same event, thus producing simultaneous faunal changes in their respective environments. The rise of the Isthmus of Panama produced convergence among the land faunas and divergence among the adjacent ocean faunas as noted in (1) and (2) above. The pattern of complementarity provides a stronger base from which to make interpretations than evidence derived exclusively from either land or marine faunas alone. (4) **Disjunct Endemism** refers to a group of fossil organisms which should have been restricted in geographical distribution due to their inherent limitations in crossing barriers to migration but which in spite of that restriction are found in widely separated areas with oceans between them. The fossil records of the reptiles **Mesosaurus** and **Lystrosaurus** discussed later provide examples of disjunct endemism.

An early example of **convergence** starts in the Cambrian (600 million years ago). The then very abundant trilobites (crab-like arthropods) are separable into two different faunal provinces which are divided by a line running through Scandinavia and Britain and eastern North America. During the next 75 million or so years of early Paleozoic time, the formerly distinct trilobite faunas became increasingly similar. Contemporaneously, several other groups of marine invertebrates, including corals, brachiopods, graptolites, etc., show the same trends. By Mid-Paleozoic time about 400 million years ago the two provinces had become one in the North Atlantic area. In chapter 8 we discuss J. Tuzo Wilson's suggestion that the boundary between the two faunal provinces was a proto-Atlantic Ocean which was narrowed by continental drift in the Paleozoic, thus destroying the deep-ocean barrier and connecting the two formerly distinct faunas. Wilson believes that the Atlantic opened twice.

A later example of Cenozoic convergence took place in the mammalian fauna of Africa which showed a strong endemic element in the hyrax (small, hare-like, hoofed mammal) and the manatee (sea cow), both found only in Africa and ancestral to the modern elephants. Such endemism supports the idea that Africa was isolated for a long time. About 25 million years ago at the beginning of the Miocene, several different mammal groups poured into Africa across land connections with Eurasia. The African fauna was reduced, and some groups became extinct; simultaneously the ancestral elephants penetrated Eurasia and rapidly attained wide distribution. The land connection between Africa and Eurasia which produced significant faunal convergence was caused by the northward drift of the African plate while rotating counterclockwise and finally colliding with Eurasia in the Miocene. (See fig. 8.13.)

The great Tethys was an ancient ocean that stretched across the region of the present Mediterranean and the Alpine-Himalayan mountain belts, lying between the northern and southern continents of the Eastern Hemisphere. During the early Cenozoic the faunas of marine invertebrates were very uniform throughout the Tethys Sea; a sharp change occurred during the Miocene epoch in which the uniformity was destroyed. The faunas of the Indian Ocean became very different from those of the Mediterranean; great divergence took place.

When Africa moved northward to collide with Eurasia in the Miocene, the connection produced convergence between the land faunas of the two continents, but simultaneously the Tethyan marine fauna was split, producing divergence in that realm. Here we are provided with an example of **complementarity.**

An example of faunal divergence is provided by Bjorn Kurten, a Finnish paleontologist who believes that the fossil record and modern distribution of reptiles and mammals were greatly affected by continental drift. The age of the reptiles, which lasted for 200 million years, includes the three periods of the Mesozoic era and the Permian period. It ended about 65 million years ago when the Cenozoic era, which extends to the present, began. Reptiles largely dominate the 200 million years of Mesozoic and Permian time, but are succeeded by the mammals, which rule during the 65-million-year Cenozoic. Kurten asks how is it that half again as many different major kinds of mammals evolved during the Cenozoic than did reptiles during the Mesozoic and Permian, which is three times longer. He believes that Laurasia, the northern supercontinent, and Gondwanaland to the south were separated by the great Tethys Sea. Although the continents may have begun to split and drift as early as the beginning of the Mesozoic (about 225 million years ago in the Triassic period), the barriers created were not great enough to restrict the movement of land animals until near the close of the Mesozoic (about the middle of the Cretaceous period 100 million years ago). This is evidenced by the fossil remains of most of the major types of reptiles which are found on both the northern and southern supercontinents. But as the continents then drifted apart during the past 75 million years in the late Cretaceous and Cenozoic eras, the barriers to mammal migrations became effective, and each of the modern continents, as fragments of the old supercontinents, became the center of and fostered its own mammal fauna through evolution during that time.

The great worldwide similarities among the reptiles on all of the continents during the late Paleozoic and Mesozoic, contrasted with the quite differentiated mammalian faunas of the more recent Cenozoic, appear to support the drift concept; but some vertebrate specialists believe that the fossil reptile evidence can be explained without moving the continents.

The evidence of sea-floor spreading indicates that North America and Europe were connected by land until the early Eocene, about 50 million years ago; apparently the connection was severed at that time. The fossil mammal faunas of the two continents support that conclusion rather nicely, because they are very similar until then but become sharply different afterward, providing an example of strong faunal divergence.

Very strong specific vertebrate evidence for drift, illustrating **disjunct endemism,** consists of the fossil remains of a two-foot-long swimming reptile named **Mesosaurus** (see fig. 4.2). It is found in the upper Dwyka beds of the Karoo rocks of South Africa's Atlantic coast just above the rocks containing the widespread **Glossopteris** flora; these beds are equivalent to the Gondwana of India. Across the Atlantic in the Parana Basin of southern Brazil is a sequence of rock beds which are also equivalent to the Gondwana and also contain the remains of **Mesosaurus.** There are great rock similarities between the **Mesosaurus-**bearing Karoo beds of Africa and the corresponding strata in which **Mesosaurus** is found in Brazil. Both are bituminous shales. **Mesosaurus** is known only from these two areas; on both continents it is found associated with fossil crustaceans and insects. Paleoecologists surmise that **Mesosaurus** was an amphibious surface-swimmer living in continental water bodies of great depth in which bituminous mud settled. Although amphibious, it had well-developed limbs and was well able to walk on land. No paleontologist believes that **Mesosaurus** could have crossed the South Atlantic; thus **Mesosaurus** supports the drift argument.

Perhaps equally strong evidence for drift came in the form of a bone fragment belonging to a Labyrinthodont amphibian; it was found in 1967 on Graphite Peak at the edge of Antarctica's South Polar Plateau. The animal was estimated at 3 to 4 feet in length and thought to have been unable to swim to the Antarctic

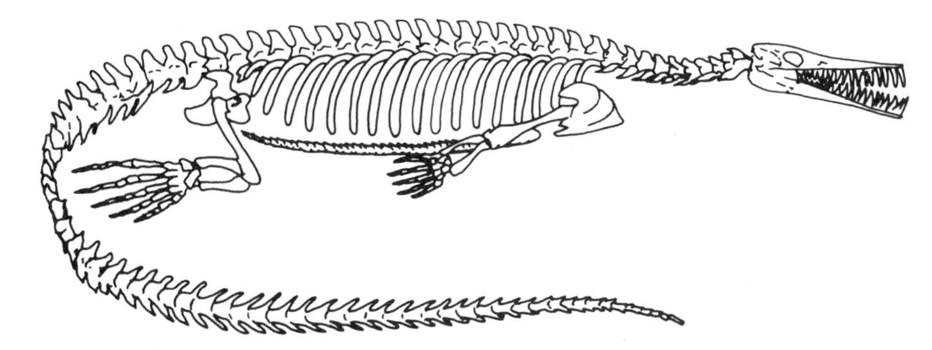

FIGURE 4.2 Mesosaurus. Skeleton of a two-foot-long, fresh-water, swimming reptile found only in the Gondwana beds of South America and Africa. It was probably not able to cross the Atlantic and therefore suggests connection of those continents.

from other continents because its family could not tolerate saltwater. Although the remains were very fragmentary and lacked strength as evidence for drift, they provided the impetus for launching another vertebrate-hunting expedition in November, 1969, led by Dr. Edwin H. Colbert of the American Museum of Natural History. In late November, after relentless digging, Jim Jensen of Brigham Young University discovered the tooth of the reptile **Lystrosaurus** (see figs. 4.3a and 4.3b) in the middle-Triassic rocks (225–180 million years ago) of Coalsack Bluff. The bluff was so named about ten years ago by New Zealanders for its coal seams up to 27 feet thick. **Lystrosaurus** was a blunt-nosed creature with only two teeth in its upper jaw; it was a short-lived species confined to the middle Triassic.

The **Lystrosaurus** tooth from Coalsack Bluff, considered together with the labyrinthodont amphibian remains from Graphite Peak, indicates that Antarctica was joined with the other southern continents in the past. **Lystrosaurus** laid shelled eggs on land — such a terrestrial reptile could not have made a

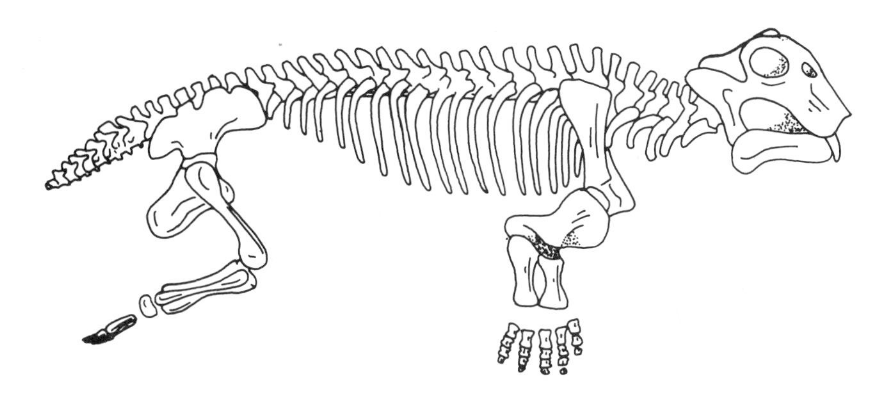

FIGURE 4.3a Skeleton of Lystrosaurus. Sketch from American Museum of Natural History collection.

transoceanic journey nor lived at the latitude in which its fossil remains are presently found. Although evidence on drift from the fossil amphibian and fish records is not definitive, perhaps least conclusive of all fossil groups to be considered in the question of drift, but worthy of mention, are the marine invertebrate faunas which are found in Paleozoic and Mesozoic sedimentary rocks of the now distantly separated continents. Many of the species belong to groups which have free-floating larvae that might have drifted across the oceans before they matured. Such mobility during their larval life stage appears to preclude them from consideration in the drift question, but it has been demonstrated that about four out of five of 200 living invertebrate species from cool, temperate waters have larval stages of less than three months and most less than six weeks. Although a substantial number of larvae of tropical species can survive an Atlantic Ocean passage in either direction, such a short larval life span for the cool-temperate-water species appears to rule out the possibility of transoceanic migrations; thus many identical, cool-water fossil species found in now widely separated continents suggest continental drift.

FIGURE 4.3b Lystrosaurus, a Land-dwelling, Egg-laying Reptile Found in Triassic-age (about 200 Million Yr Old) Rocks at Coalsack Bluff in Antarctica. It does not appear that it could have lived at the latitude of its fossil remains and therefore supports continental drift, movement of the pole, or both. (Courtesy of Edwin H. Colbert.)

A coral is an animal much like an inverted jellyfish which builds a rock-hard, limy house. Some corals grow together in great colonies to build reefs; fossil reefs have a great bearing on continental drift since modern, growing coral reefs are largely confined to the tropics and thus may aid in delimiting ancient environments. Such corals lived well within the Arctic Circle in the Canadian Arctic islands and Greenland during the early Paleozoic era over 350 million years ago. It appears that the wide zone containing fossil reef corals has moved southward toward the modern tropical zone since that time. If ancient reef corals were confined to the climatic zones required by their modern representatives, then this implies that both Europe and North America have drifted a considerable distance to the north. It really does not seem probable that the Arctic of the Paleozoic seas was warm enough to support reef-building corals while the tropical seas were too warm for them to thrive. Nor does it appear likely that the reef-building corals of the Paleozoic prospered in cold Arctic seas.

P. M. S. Blackett believes that the average latitude of the fossil coral reefs of any given geological period probably represents the position of the equator at that time since the modern ones are roughly centered about the equator. He summarized his data in table 4.1.

TABLE 4.1

	Range of Age in Millions of Yr	Mean Age in Millions of Yr	Mean Latitude of Reef Coral
Cretaceous Jurassic Triassic	70–225	150	35°N
Permian Carboniferous	225–350	290	61°N
Devonian Silurian	350–440	395	47°N
Ordovician	440–500	470	61°N

Calcareous algae are single-celled seaweeds that build rocky structures of lime (calcium carbonate). In many places in the shallow seas we find large, banded, pillow-like masses of these calcareous, algae that are forming reefs. In living algal reefs, the algal structures are turned toward the direction of maximum light from the sun because algae are similar to all plants in their heliotropism, or positive response in turning toward light. Fossil algal reefs dating back 1200 to 600 million years from near Peking, China, show that the convex upper surfaces of the banded, algal pillow structures are similar to living forms. Careful measurement of the direction in which the surface convex bands point indicates a progressive change in the angle at which maximum sunlight was striking the algae. If the angle of the sun's rays were shifting gradually during this 600-million-year interval, it could mean that this region of China has moved southward 40° in latitude. It is possible instead that the North Pole wandered to the north from its formerly more southerly position by that same amount of latitude. It is now generally believed that both the land and the pole have changed positions.

Not all invertebrate evidence favors drift. F. G. Stehli of Western Reserve University, and others, have shown invertebrate evidence contrary to continental drift and polar wandering. Using the principle, derived from studies of modern faunas, that the greatest numbers of different kinds of organisms occur near the equator and fewer and fewer different species are found as we move toward the poles, they concluded that fossils from the Permian and other periods indicate that diversity distributions in the past have been almost paral-

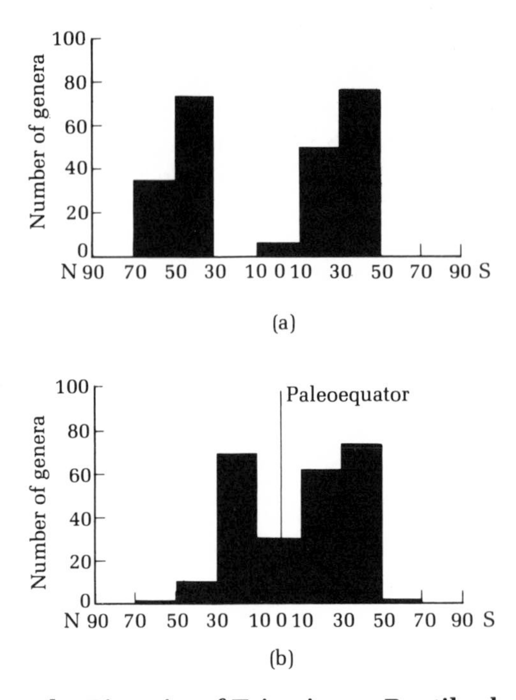

FIGURE 4.4 The Diversity of Triassic-age Reptiles by Number of Genera Plotted according to the Latitudes of the Fossil Localities. Fig. (a) shows plot for modern latitudes. Fig. (b) is based on Triassic-age latitudes for fossil sites as interpreted from paleomagnetism studies. (After D. A. Brown, *Australian Journal of Science* 30 (1968): 439. Reprinted by permission.)

lel to the present equator. They maintain that such a pattern of diversity for fossil species seems to speak for stable continents.

In contrast, D. A. Brown has noted that the distribution of reptiles and amphibia of the late Paleozoic and early Mesozoic from all of the continents does not show the greatest diversity near the equator. This diversity distribution is also not symmetrical about the present equator. He summarized the data concerned with the distribution of different kinds (genera) of reptiles and amphibia and concluded that they provide strong evidence that those fossil animals were not buried at their present latitude in late Paleozoic time. (See fig. 4.4.) It appears that they have drifted considerable distances since their entombment over 100 million years ago.

Many other studies might be reviewed, but as we summarize the arguments for and against drift and polar wandering drawn from the total fossil record, we are forced to conclude that they are equivocal and nowhere really definitive. Many members of the paleontologic community have come to the view that geologic, rather than fossil, evidence will really decide the drift question. The fossil evidence still appears to lack fine definition. Questions about dispersal of organisms, the great adaptability and tolerance of fossil groups, the possibility of ancient land bridges, and other arguments have kept paleontology from a definitive role in the argument since Wegener's presentation.

Ancient Climates

Paleoclimatology, or the study of ancient climates, was a major line of evidence presented by Wegener in his original work of 1915. He used it in the reassembly of the protocontinent of Pangaea and in location of ancient pole positions. The rock formations of the world, by their character and distribution, offer eloquent testimony about the climates in which they were deposited. The former presence of tropical forests in what is now the Arctic, the hot-desert deposits in areas presently receiving 80 inches of rainfall annually, and the ancient glacial deposits located on the equator may be explained in more than one way, but all explanations use the principle of **uniformitarianism.** This principle assumes that the evidence of the rock and fossil record came into being throughout the geologic past by the operation of the very same forces and processes which presently operate on the earth. If we view such evidence which is not in keeping with the present climatic zonation of the earth, two possibilities become most apparent: either the continents have drifted great distances across major climatic zones, or the climatic zones have shifted greatly because of major changes in the forces that control them.

The elevation of the sun above the horizon at noon determines the latitude, or distance north or south of the equator. The more nearly vertical the sun's rays, the greater the heat energy received. The earth's climatic zones are controlled mainly by this fact. The three main zones are the **polar** (in which the sun is below the horizon for at least one day per year), the **temperate** (in which the sun is always less than 90°), and the **tropical** (where the sun is directly overhead sometime each year). (See fig. 4.5.) The position of the sun accounts for the fact that the equator receives five times as much solar heat as the land surface at 80° north or south latitude. Local variations in the zones are produced by the size and shape of sea masses and landmasses, the presence of mountains, plateaus, and valleys, and the movement of currents within the oceans and seas.

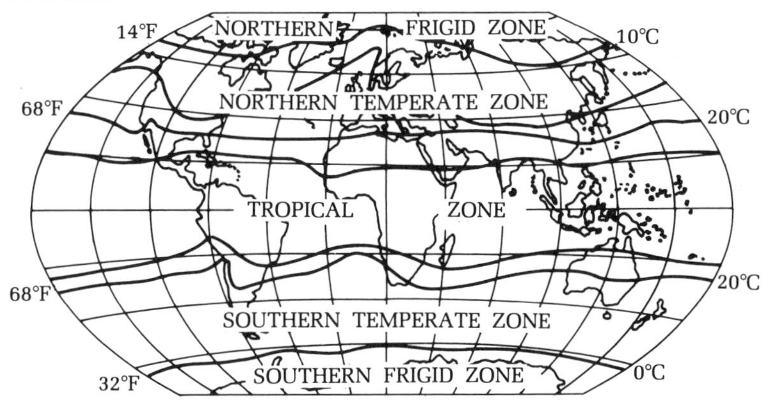

FIGURE 4.5 Climatic Zones of the Present. The zones are closely related to latitude and are controlled mainly by the angle at which the rays of the sun strike the earth's surface. The irregularities in the zonal boundaries are due to the effects of the distribution of water- and landmasses and the action of major cold- and warm-water ocean currents.

In order to better understand the climatic evidence from the geologic record, let us first look briefly at the present conditions of climatic zonation. Solar heating in the tropical zone generates moist, low-pressure, rising air which yields rain. This dried air, after dropping its moisture, moves toward the poles at great elevation, curling inward to become compressed, and drops as dry air in the Horse latitude belts. Dry, cold, heavy air descends over the poles and flows outward to form a low-pressure polar front along the border with the temperate zone. The dry, high-pressure zones are constant from season to season but vary in strength, but the low-pressure belts, especially along the polar fronts are highly variable with spotty rainfall and variable winds.

The lands of the middle latitudes warm and cool more rapidly than the adjacent water bodies, causing monsoonal air flow which breaks up the Horse latitude belt of high pressure. Monsoon centers are dry with great evaporation of surface waters, often producing sulphate and carbonate (salt) deposits and playa (intermittent) lakes.

The climates produced by the movement of the major air masses are familiar to all. The dry, falling air moves toward the equator from the Horse latitudes and the monsoon center and dries the lands to deserts or near deserts. Little rain supports only succulents and annual plants, and old bays and lagoons cut off from the sea evaporate, yielding salt beds. Rising air at the equator yields much moisture to support lush forests. Disturbances along the polar front produce frequent variations in moisture and temperature in the temperate zones, along with the peculiar plants and animals of such regions.

The trade winds between the Horse latitudes and the equator promote a westerly movement in the ocean. The ocean drift continues until striking a landmass — most of the water continues westerly to return to its starting point, but some escapes into polar seas to be chilled and return as a polar current.

Low-elevation air circulation is broken up by mountains and varies greatly over short distances. Warm, humid air is derived from the warm ocean currents; cool, dry air comes from the cold regions. Deserts occur more often along the cold currents of the west coasts than those of the east coasts, and zonation is more strongly developed in the western sides.

In examining the past, we may expect with a certain confidence that the greatest changes in climate would have been produced by changes in latitude, rather than by uplift of mountains and reduction of landmasses by erosion.

The geologic record contains very strong proof that the continents of the past have been subject to climates radically different from those of the present. We shall discuss several lines of evidence, but let us begin with Wegener's most important one — glacial evidence.

There are clearcut signs that about 300 million years ago (Pennsylvanian-Permian periods) great masses of ice in the form of continental glaciers moved across the southern continents, scouring the rocks under them and leaving distinctive glacial deposits behind. These deposits form nonsorted, nonstratified, very mixed sedimentary rock called **tillite** — irrefutable proof of the former presence of glaciers. In 1856, Blandford discovered such a tillite in tropical India at Talchir in Orissa. In 1859, tillites of the same age were found in South Australia; in 1870, such tillites of the same age were mapped also in South Africa. Eighteen years later they were discovered in Brazil. The glaciers necessary to the formation of tillites in the southern continents (which then formed the so-called protocontinent of Gondwanaland) were operating at the same time that most of Laurasia (North America, Europe, and Asia, excluding Arabia and Peninsular India) was under tropical or hot-desert climates. This is truly paradoxical in terms of present climatic zones! The glaciation persisted for about 50 million years at some localities and produced deposits up to 300 feet thick and covering hundreds of thousands of miles in some cases.

Additionally, it is often quite simple to determine the direction of movement of glaciers from the scratch marks or striations produced as the glacier pulled embedded boulders and cobbles across the exposed bedrock over which it advanced. The boulders and debris picked up and carried great distances can often be traced to their place of origin and provide another clue to the direction of ice movement.

All such evidence suggests that the glaciers moved westward across Africa out into the present Atlantic. The glaciers passed northwestward on to South America. The glaciers of Brazil carried rock types found in southwest Africa but not found elsewhere in South America. There are many other localities with similar glacial evidence from the other southern continents which requires that we reassemble the southern continents into a single landmass or somehow miraculously explain how glaciers originated in what are now tropical, temperate oceans and picked up glacial till which includes continental-type, igneous rocks and carried them onto the southern continents.

J. C. Crowell and L. A. Frakes (1970) have done a detailed study of the distribution of late Paleozoic (350–225 million years ago) glacial deposits of Antarctica, Australia, South Africa, and South America. They show that glaciers grew at and radiated from centers which appear to coincide approximately with the site of the earth's magnetic pole at that time; thus the glacial and paleomagnetic evidence (discussed in chapter 6) is in agreement.

The hypothesis of the origin of glaciers which appears most in keeping with the available evidence was advanced by M. Ewing and W. L. Donn. They believe that ice caps begin to grow when the earth's rotational poles move onto the land or onto landlocked water bodies. Glaciers could not grow if the poles were situated in mid-ocean. They maintain that the most recent ice age, the Pleistocene, was triggered when the south pole moved onto the Antarctic landmass and the north pole moved into the Arctic Ocean which is nearly surrounded by land. Thus we should expect that large landmasses be concentrated at the poles during major glacial periods. All other lines of evidence suggest that the great southern protocontinent of Gondwanaland was in a south polar position during the late Paleozoic when its great glacial deposits were formed and thus appears in keeping with the Ewing-Donn hypothesis.

We may now turn to a quite different kind of evidence to show that while the continents of the Southern Hemisphere were under ice in large part during the Permian and Carboniferous periods, those of the Northern Hemisphere were developing extensive evaporate deposits. Evaporites are rocks which form in bodies of water in which the rate of evaporation is greater than the rate of flow into the basin. The most common evaporites are salt and gypsum. Ideal conditions for the development of evaporites include a hot and arid climate. If we now recall the present climatic zonation of the earth presented earlier, we find that such conditions do not exist in the latitudes at which we find the great evaporite deposits of the Permian and Carboniferous periods. The greatest of all mapped salt deposits are of Permian age in Europe and the U.S. Great deposits of evaporites lie across the southern United States, Britain, Germany, the southeastern section of the Russian Platform, and the western flank of the Ural Mountains. (See fig. 4.6.)

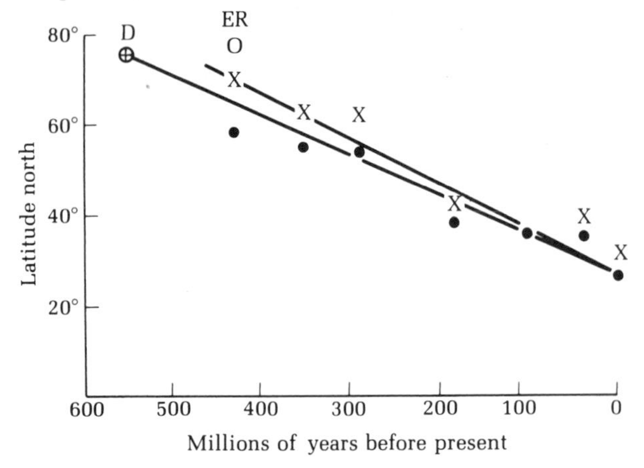

FIGURE 4.6 The Distribution of Major Salt Deposits of the Northern Hemisphere. Present average latitudes are plotted against ages. X, Europe-Africa; •, North America; ⊕ D, Devon Island, Cambrian age; O ER, Ellef Ringnes Island, Silurian age. (Modified after Holmes, 1965.)

It is believed that the great evaporation which formed these deposits must have taken place at approximately 15° to 40° latitude during the Permian period (about 250 million years ago). The western flank deposits of the Ural Mountains are located at 60° north latitude — an unlikely climatic zone for their formation. Eurasia seems to have drifted at least 25° to the north in about 250 million years since the Permian period. Great salt domes of Cambrian to Silurian age in the Queen Elizabeth Islands of Canada at about 77°N latitude suggest a northerly drift here of about 40° in about 450 million years. The evaporite evidence is not precise enough to define the exact position of the continents in past periods, but examples such as these and numerous other salt deposits of various geological ages in America, Africa, and Europe generally point to a northerly drift of these continents, since the salt was formed. What is suggested by evaporites agrees closely with the evidence from ancient rock magnetism which we shall examine later in chapter 5.

The study of ancient wind directions can be done by examination of wind-deposited, or aeolian, sandstones. The cross-bedding produced within a dune is a reliable guide to the direction of the wind which formed it. (See fig. 4.7.) Studies of modern dunes indicate that the direction of dune movement can be expected generally to agree with the general wind pattern of the earth, except under monsoonal conditions found on the eastern coasts of the major continents. Opdyke and Runcorn and others have indicated that wind-deposited, Upper Paleozoic sandstones of Europe and North America show directions suggesting a trade-wind belt which agrees with the reconstruction of the equator of that time based on ancient rock magnetism. The method of determining the direction of ancient winds was originated by F. W. Shotton in his 1937 interpretation of the wind-deposited, Permian-age New Red Sandstone of Britain. His study indicated that desert sands in various parts of Great Britain had been formed by northeast trade winds during the Permian. This means that Britain was within about 30° of the equator; thus it has moved northward about 25° and also has rotated about 40°. (See fig. 4.8.) Unfortunately, this technique is limited by the infrequent occurrences of such aeolian sandstone.

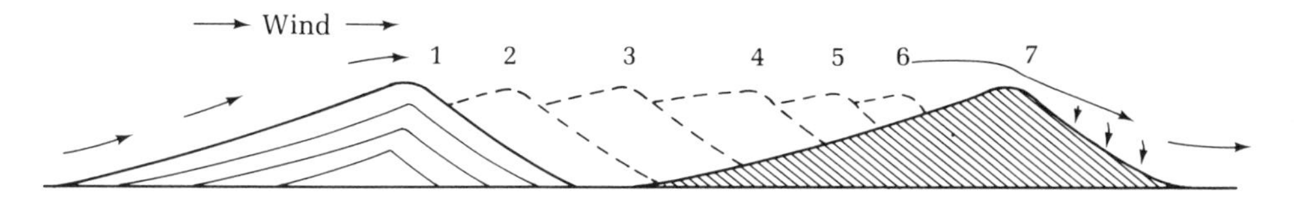

FIGURE 4.7 Series of Cross Sections to Illustrate the Growth, Movement, and Bedding of Sand Dunes. A stationary dune at (1) increases in height with an upward and forward advance of its crest. If the wind velocity and supply of sand reach certain limits, the dune migrates coming to occupy positions (1) through (7) successively. The slope of the bedding within the dune structure indicates the direction of prevailing wind.

FIGURE 4.8 Arrows Indicate Direction of Dune Advance in the Permian-age Desert of Britain. The map has been turned so that the Northeast trade winds of the Permian period agree with the directions indicated by the cross-bedding in the wind-deposited sandstones. We then infer that Britain was about 15°–30° north of the equator and has since drifted northward and rotated about 40°. (After F. W. Shotton and S. K. Runcorn *fide* A. Holmes, 1965.)

Another method useful in ancient temperature analysis involves the use of a mass spectrometer for measuring oxygen isotopes. Isotopes are like the other atoms of an element but have a different atomic weight. Oxygen 18 and oxygen 16, two different isotopes, occur in limy shells of sea animals in varying proportions according to the water temperature. The oxygen 18 increases as the temperature falls. Long study has revealed that the ratio between the two isotopes enables us to make temperature estimates to within one-half degree Centigrade (approximately one degree Fahrenheit). Assuming that ancient marine life and chemical processes operated then as today, such studies done on Jurassic-age (about 150 million years old) fossil, shelled relatives of the squid (belemnites) from rocks at 57° north latitude indicate that the average annual seawater temperature in which those Jurassic, squid-like animals lived was roughly 15° Centigrade warmer (14°–20°C) than presently prevails at that latitude.

The Biological Effects of Continental Drift and Reversals of the Earth's Magnetic Poles

The fossil record provides eloquent testimony about the rise and sometimes sudden fall of diverse groups of plants and animals. Worldwide disturbances of habitats have resulted in mass extinctions of major groups of organisms including the dinosaurs, plants, marine invertebrates, and many others. Earth history is punctuated by major and minor episodes of widespread extinctions. Until the nineteenth century the great discontinuities in the sedimentary rock record with their accompanying breaks in the fossil record were attributed to worldwide catastrophies likened to the biblical flood. Since the breaks in the fossil record seemed to be repetitive, theorists of the late nineteenth and twentieth centuries turned to the idea of a cyclic or rhythmic force within the earth to explain repeated destructions of environments. They referred to movements of the earth's crust that could produce changes in the configuration of oceans and continents and great alterations of climate. The evidence for such repetitive change on earth lay well documented in the rock and fossil record, but the engine, the basic cause, was not understood.

We now possibly have such an engine — convection currents in the mantle repeatedly moving large plates of the earth's crust away from and toward each other through geologic time may have produced such seemingly "cyclic" changes. If there have been major variations in the rate at which crustal spreading and volcanic eruptions of lava have taken place in the mid-oceanic ridge areas, then such motions of the oceanic crust must have changed the volume of the ocean basins. Disturbances which altered the configuration of continents and water bodies also must have greatly affected climates and led to major migrations, extinctions, and accelerated evolution of many animal and plant groups. If such disturbances of oceanic and terrestrial living space have taken place repeatedly through geologic time, it is quite possible that as our knowledge of the details of continental drift and sea-floor spreading expand, we may well be able to correlate the possibly cyclic character of such crustal movements with changes in the fossil record.

It appears that there may be connections among periods of volcanic activity, continental drift due to the spreading of the sea floor, and the frequency and rate at which the magnetic north and south poles have reversed themselves through geologic time. The evidence for repeated magnetic polar reversals is discussed more fully in a later chapter.

The cause of the earth's magnetic field with its magnetic poles north and south is still theoretical. Recall that the earth has a solid inner core surrounded by a liquid outer core about 1300 miles thick. The mantle surrounding the liquid outer core is about 1800 miles thick. The crust on top of the mantle is 3 to 35 miles thick. Most geophysicists now believe that motions in the liquid outer core due to the earth's rotation are mainly responsible for the earth's magnetic field. This theory is supported by the facts that Jupiter, which rotates every ten hours, has a stronger magnetic field than the earth, that the moon rotates only once every 27 days and has a very weak magnetic field, and that Venus, which rotates even more slowly than the moon, has no magnetic field.

During the interval of magnetic polar reversal, the earth's magnetic field is believed to have collapsed, resulting in the destruction of the Van Allen belt. The Van Allen radiation belt, discovered in 1958, consists of two doughnut-shaped rings (see fig. 4.9). The inner ring lies 2000 miles away from the earth and the outer is about 10,000 miles distant.

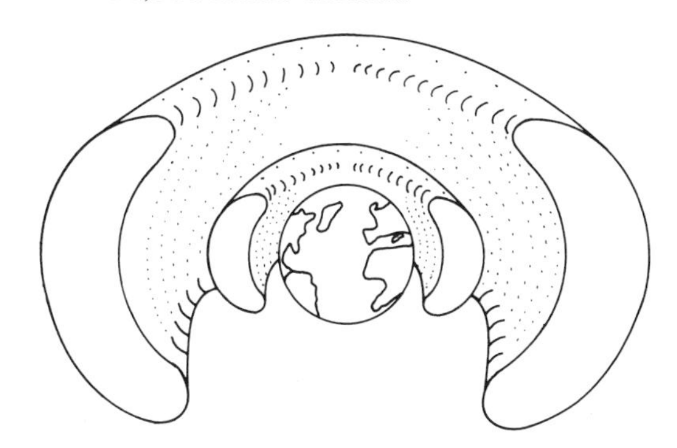

FIGURE 4.9 The Van Allen Radiation Belts. The inner ring lies 2000 miles from earth; the outer ring is about 10,000 miles distant.

The particles in the outer belt, trapped in its magnetic field, are protons and electrons derived from split atoms; they probably come from the sun. The particles in the inner belt appear to come from the earth's atmosphere. The universe has been called a shooting gallery of electrically charged particles called **cosmic "rays."** They are composed of the nuclei of atoms which have had their electrons removed. Most of the cosmic "ray" particles are deflected away from the earth by the earth's magnetic field; others are trapped in the Van Allen radiation belt, but many particles strike the earth. Cosmic rays are strongest at the poles and weakest at the equator since the charged particles travel along the magnetic field lines but tend not to cross the magnetic field. All life on the earth is being struck constantly by cosmic rays which produce genetic changes in organisms, thus affecting the evolution of whole populations. Changes in the earth's magnetic field and especially polar reversals with collapse of the field and Van Allen belts may have played crucial roles in the evolution of life and possibly led to extinctions of whole groups of organisms. Such changes appear to have been linked additionally to volcanic activity which probably affected the atmospheric dust levels; the dust levels in turn would have influenced precipitation and solar effects on the earth. It has been further suggested that faunal extinctions and polar reversals may be the mutual results of a third, indirect cause; tektites are small (marble-sized), black to green or yellowish, rounded masses of silicate glass. They have various shapes, indicating an airborne origin, and are thought to be extraterrestrial or formed by meteorite impact on earth rocks. A great rain of tektites took place across parts of Asia and Australia about 0.7 million years ago, which coincides with the time of the last geomagnetic polarity reversal. Perhaps the meteor impact which formed the tektites wiped out large parts of the surface-dwelling marine fauna and jolted the earth's core to produce a simultaneous reversal of the magnetic poles. The exact relationships between plate tectonics, continental drift, sea-floor spreading, and reversals of the earth's magnetic field and volcanism are not yet understood, but there appear to be some coincidences of periods of increased change in the fossil record and the dates of the magnetic polar reversals.

If a large landmass is pulled apart, it results in a great increase in coastline area — especially important is the increase in the continental shelf and slope areas, which are shallow parts of the sea bordering the continents. Sea life is most abundant and diverse here. Such an increase in space well-suited to invasion and occupancy by many life groups would undoubtedly contribute to an increase in the number of different species.

The climates of landmasses are greatly influenced by the size and location of nearby water bodies. Additionally, currents in the oceans produce marked effects on climate. The splitting and separation of a landmass with the creation of a water body between the segments would profoundly affect their climates. Australia was a part of the Antarctic continent until about 50 million years ago in the Eocene, when it separated and began its northward drift, providing a water area for the creation of the Antarctic current, which must have greatly altered the climate. As Australia moved northward toward the equator, its climate became increasingly warm and hospitable to allow the evolutionary development of diverse plant and animal groups which certainly were not likely to develop in the Arctic.

The separation of areas which were formerly joined would lead to discontinuous distribution; as separation continued, the animals and plants in the two areas would become less and less similar through evolution. After long-term separation, perhaps there would be few species, genera, or even families in common. If any of the areas remained isolated for great lengths of time, we might expect that plant and animal types peculiar to the region would develop; such endemic organisms are, in themselves, an indication of long-term isolation.

In the movement of huge plates of earth crust, continents are brought together as well as carried apart. Many such movements have produced crashes between continents which have resulted in the creation of mountain ranges along the colliding continental fronts. The Appalachians, Alps, Himalayas, Urals, and many other major mountain ranges are believed to be the product of such collisions. Australia is presently drifting into Asia; India's collision with southern Asia during the past 60 or so million years seems to have produced the Himalayas.

Just as continental separations would affect climates by the creation of water bodies, collisions would destroy water areas and produce quite different climatic effects. The raising of a major mountain range would greatly alter wind patterns, precipitation, and temperatures and would reduce the living space for various life groups. Such environmental changes produced by collisions would, as in the case of continental separations, put great pressures on the inhabitants and result in increased extinctions and accelerated evolution.

Norman D. Newell has recently shown the impact of plate tectonics in the interpretation of the community of marine plants and animals that build tropical reefs or "coral reefs." He correlates the major changes in reef communities with major events in earth history. Tropical reefs consist mainly of fine, sandy sediment bound together by the plants and animals fixed to its surface. Corals are the most commonly known of the reef organisms, but many diverse groups contribute to its growth, including lime-secreting protozoans, relatives of the sand dollar, clams and oysters, and many others. The modern reef community lives in warm water with an average winter temperature between 27° and 29°C in regions without marked seasonal changes. The reef grows best in clear, well-oxygenated water of normal salinity; its community is described as "remarkably sensitive to environmental change." The paleontologist applies the principle of uniformitarianism in his inference that fossil reef communities composed of organisms similar to their living descendants must have lived under conditions similar to those of modern reefs. Thus, profound changes in fossil reef communities must have been wrought by major environmental change. Newell believes that the record of fossil reefs clearly marks a number of "catastrophic episodes" in earth history.

Although periods of marked extinction of many fossil groups occurred earlier, at the end of the Paleozoic era (about 225 million years ago), about one-half of all the families of both sea and land animals and many land plants became extinct. It is thought that all the continents came together at that time to form Pangaea, the great protocontinent. Glaciers scoured the southern Gondwana continents during those times of cooler climates, and the reduction in ocean basin water volume due to glacial ice growth probably lowered sea level and greatly increased the land area. Such a reduction in ocean surface area must have contributed greatly to climatic extremes and thus led to major extinctions including members of the Paleozoic reef communities.

Following that great extinction time, many new groups arose and reef communities flourished around the world, with some old members and many newly evolved groups. During the Mesozoic era, life thrived and expanded, and the seas and land were richly endowed by a diversity of animal genera that reached its peak in the Cretaceous period. At the end of the Cretaceous, extinctions took place on a gigantic scale. Almost one-third of all known Cretaceous families of animals failed to survive into the next era, the Cenozoic. The ammonites, relatives of the pearly nautilus, became extinct; two-thirds of all coral genera died out. The highly successful clam-related rudists, which had become important Cretaceous reef builders, disappeared completely. The Mesozoic lands had been dominated by the dinosaurs, but not a single one of the 115

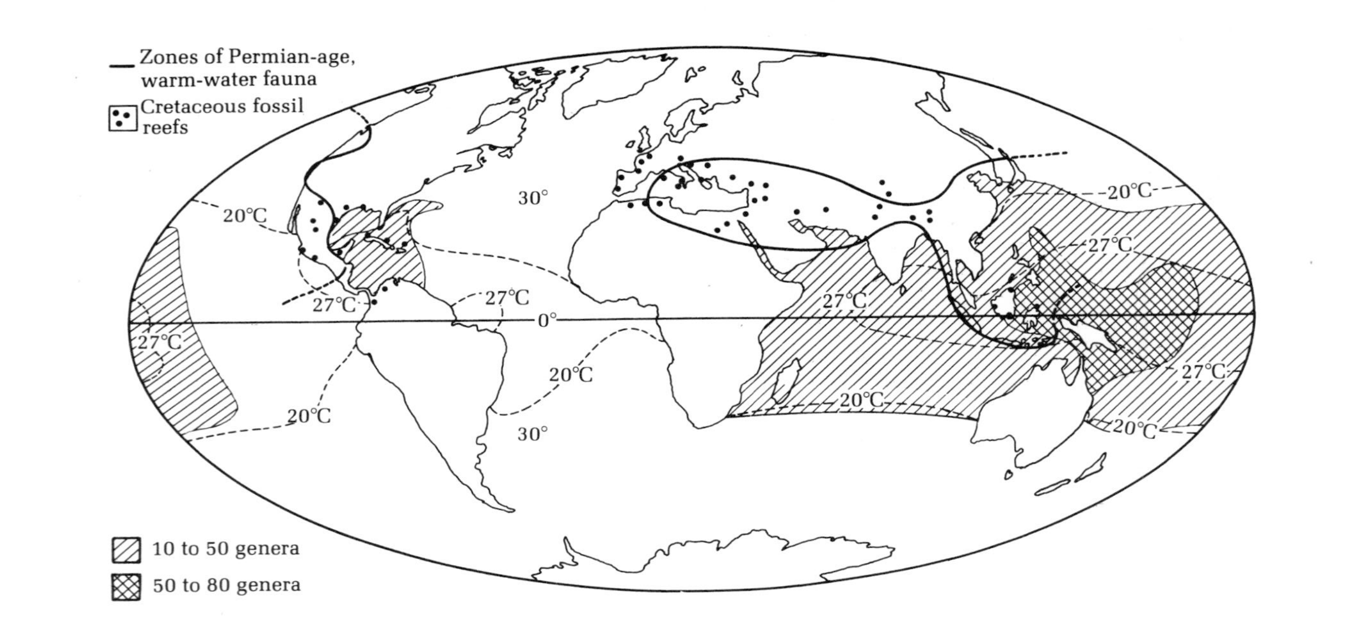

FIGURE 4.10 Modern Reef-building Organisms. They live in a zone about 60° wide centered on the equator (lined area). Greatest species diversity occurs where the winter average water temperature is 27°C. Reef fossils of Paleozoic and Mesozoic age are found far north of modern coral reef limits, indicating that the ancient equator was located far north of the present one. The asymmetrical distribution of the fossil reef zones compared to the present one suggests to Newell that large parts of once more widely distributed fossil belts have been destroyed by subduction along plate edges. (From "The Evolution of Reefs," Norman D. Newell.

different Cretaceous genera survived the catastrophic environmental change at the end of the Mesozoic era.

Newell and others believe that the Mesozoic world was "predominantly a water world which fostered a broad zone of gentle climate around the equator." Most of geologic time has seen much less land than the 29% of the global surface that exists at present; an increase in land area contributes to strong ocean currents, strong winds, and extremes in seasonal climatic variations. Figures 4.11a and 4.11b show the apparent correlation between the diversity and decline of animal genera at the end of the Cretaceous and the reduction of water area on earth at that time.

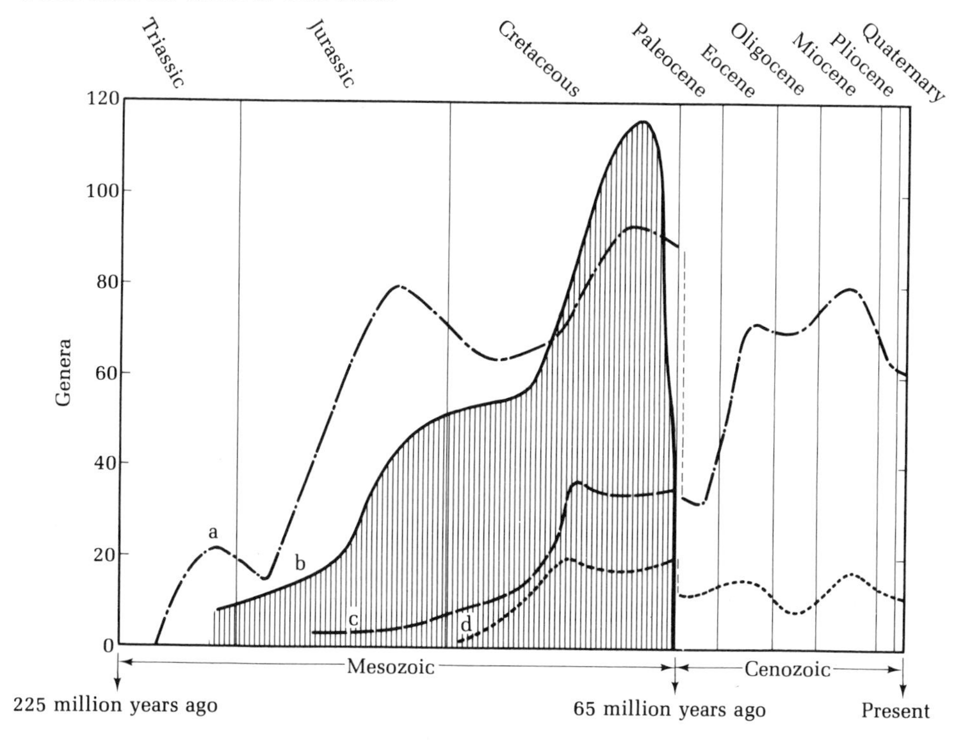

FIGURE 4.11a Diversity and Decline of Animal Genera. Mesozoic variety of land and ocean genera peaked in the late Cretaceous and was followed by mass extinctions at the end of that period. Lines (a)–(d) trace the diversity of four animal groups by genera showing a marked change for all at that time about 65 million years ago. (a) The corals declined; (b) the dinosaurs became totally extinct; (c) the clam-related rudists also disappeared; and (d) the free-floating, miscroscopic, globigerine foraminifera were greatly reduced. Compare with fig. 4.11b. (From "The Evolution of Reefs," Norman D. Newell. Copyright © June 1972 by *Scientific American*, Inc. All rights reserved.)

The joining of two major landmasses, through drift, each with its own groups of plants and animals, would lead to increased competition among many similarly adapted forms thrown into competition. The eventual outcome of such a

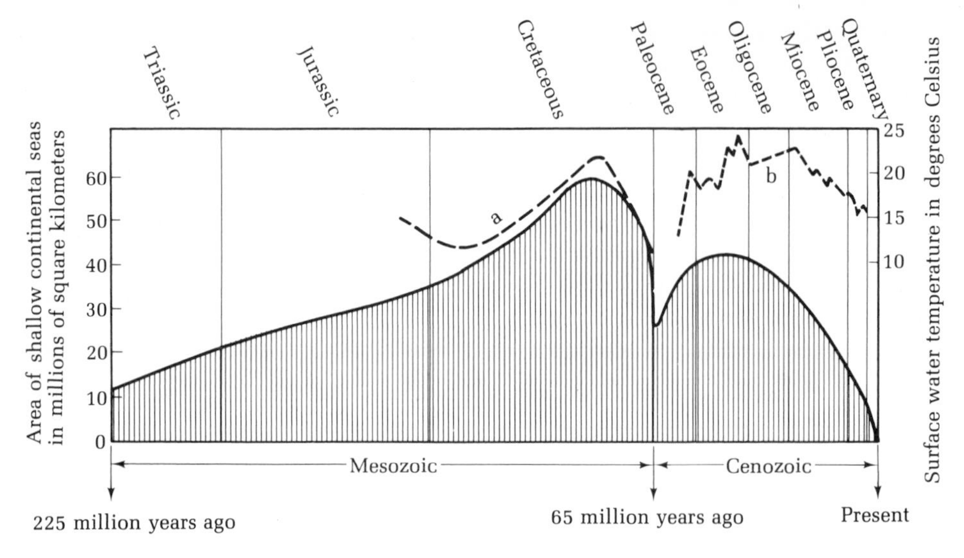

FIGURE 4.11b Reduction in Area of Shallow Continental Seas and Surface Water Temperatures during the Mesozoic and Cenozoic. Near the end of the Mesozoic, about 85 million yr ago, the seas began to shrink in surface area; the shallow regions were most affected. The temperature of the seawater is indicated by oxygen isotopes found in the skeletons of belemnites (a) which are extinct relatives of the squid and microscopic foraminifers (b). A cooling trend accompanied the reduction in the area of the shallow seas in both the Cretaceous and Cenozoic. Compare these curves with those for animal diversity in fig. 4.11a. (Based on Emiliani, Lowenstam, Epstein, Durham, Newell, and others.)

continental union would be a reduction in the diversity of species. Such landmass collisions, once understood, would play central roles in deciphering the migration and dispersal patterns of plant and animal groups through space and time by the biogeographer and paleobiogeographer. J. W. Valentine and E. M. Moores (1970) have shown that the diversity of invertebrate families may be correlated with plate tectonic patterns. The number of families increased from about 75 in the Cambrian (600 million years ago) to almost 300 in the middle Ordovician (450 million years ago) following the breakup of the earth's single protocontinent. The number of families fell sharply in the late Permian (240 million years ago) to just over 100; this decline is correlated with the reassembly of the continents at that time which reduced the continental shelf space. From the Triassic (220 million years ago) to the present there is an almost steady, strong increase to about 450 families; this most recent increase in variety of invertebrate types they relate to the breakup of Pangaea which resulted in greatly expanded continental shelf space for the evolution of new groups. D. M. Raup (1972) has reviewed the work of Valentine and Moores, and Newell, on the diversity of invertebrate families and also Müller's diversity plots of invertebrate genera. Raup suggests that "changes in the quantity of the sedimentary rock record may cause changes in apparent diversity by introducing a sampling bias." He suggests caution in interpreting diversity data and thus detracts from faunal diversity arguments related to drift and plate tectonics. Such profound questions will surely evoke further studies and more complete answers.

If continents have in fact collided and separated in the past, providing for the intercontinental exchange of then living passengers, then we might also expect that the fossils entombed in the sedimentary rocks of one continent have in some cases been "plastered" onto the other and left behind. Such possibilities have never before figured seriously in biogeographic and paleontologic interpretations (McKenna, 1973).

We have come to a new view of the changing animal and plant world through geologic time; many old questions about extinctions, migrations, and variations in rates of evolution are destined to yield to the newly discovered facts of plate tectonics, continental drift, and ancient geomagnetic changes.

5

Rising and Sinking Crust, Heat from the Earth, and Currents in the Mantle

Isostasy

If mountains were merely high masses of rock sitting on top of a rigid uniform crust, then we should expect to find the force of gravity greater on top of the mountain than on the surrounding lowlands. This is so because after the difference in altitude is corrected, the greater mass of rock making up the mountain would add to the gravitational pull. Surveys in the middle 1800s in India by the British show that this is not the case, and it is now widely believed that the surface irregularities of the earth are not supported by the strength of an unyielding crust; the crust appears instead to be composed of many small segments floating on a dense plastic interior below it. (See fig. 5.1.)

The state of flotational balance of the crust is called **isostasy.** Such a theoretical flotational state, or isostasy, could be attained in one of two ways or a combination of both. For example, if we floated an oak and a balsa-wood plank in a tub of water, the balsa would float much higher above the water surface than the oak — thus representing the mountain mass (see fig. 5.2).

A second possible explanation assumes that the rocks of the entire crust, including mountains, plains, plateaus, ocean floors, etc., are all of uniform density but that they rest on a denser, plastic layer beneath them. In this case the highs and lows of the earth's surface result from blocks of crust which differ in thickness, not density (see fig. 5.3).

We now know that the earth's crust is not divided into simple blocks which slide freely past each other such as those in figures 5.2 and 5.3. We have also learned that the upper mantle (the layer beneath the crust) is not a fluid but may flow plastically under heavy loads over a long period. It was formerly believed, and incorporated in various theories on isostasy and movements of the earth's crust, that the important boundary to consider was the Moho, or Mohorovičić discontinuity, which separates the crust of the earth from the mantle. About ten

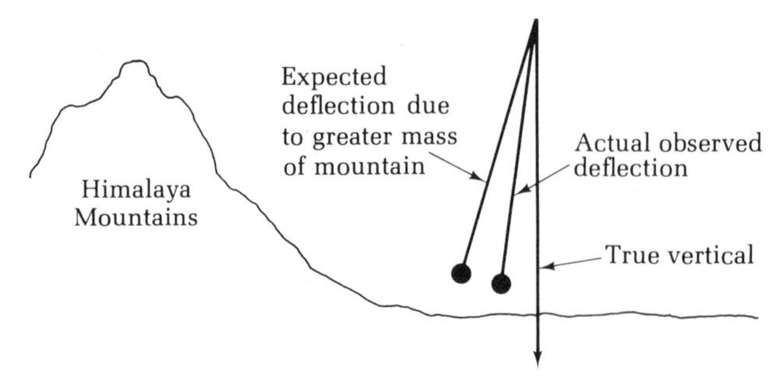

FIGURE 5.1 Early Evidence for Isostasy. British surveyors of the last century expected the greater local mass of the Himalaya Mountains to attract the plumb bob and deflect it greatly from the vertical position. Instead, the deflection was considerably less than expected, indicating that light rocks underlie the mountains.

years ago Don L. Anderson refocused attention on a more important, poorly defined boundary between the rigid lithosphere and the weaker asthenosphere. He believes that most of the movement within the mantle takes place in the plastic layer at the top of the asthenosphere. Discovery of this plastic layer which extends from approximately 40 to 150 miles in depth contributed greatly to the plausibility of convection current theory.

The upper mantle immediately beneath the crust must behave something like sealing wax, which breaks like a solid from a sharp blow but bends plastically or sags under slowly applied stress such as that which it suffers if left unsupported. The earth's crust is very strong, and great weight must be imposed on it before it begins to yield by plastic flow. If the various parts of the crust are floating on the mantle like rafts on water, then there must be a depth below which the pressure is equalized — the blocks of crust do not sink below this depth, which is called the **compensation depth.**

The mountains are made of lower density rock and therefore float higher than do the ocean floor rocks which are of greater density. The level at which pressure is equalized (the compensation depth) is approximately 35 miles deep. Most mountains are no higher than about 20,000 feet because that is about as high as they can stand within the system of flotational balance (isos-

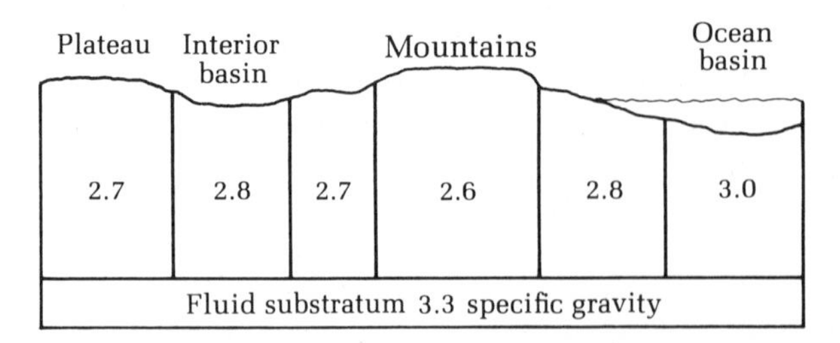

FIGURE 5.2 Pratt's Isostasy Theory Modified Slightly to Show the Mass of Different Blocks of Crust Based on Modern Estimates. Numbers are specific gravity (weight of a substance as compared to an equal volume of water). The block forming the mountains is 2.6 times heavier than an equal sized block of water.

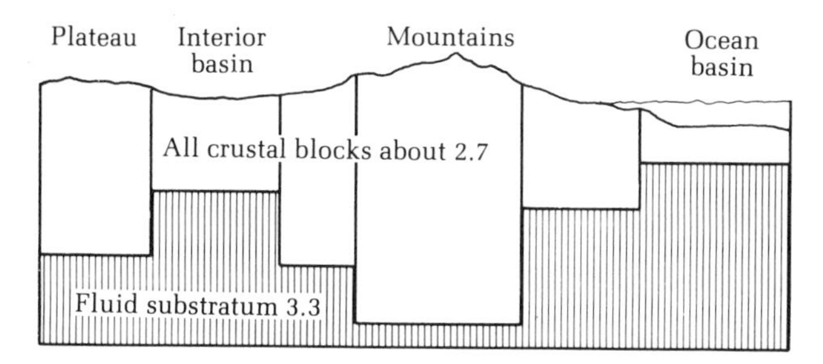

FIGURE 5.3 The Mountain Roots Theory of Isostasy Proposed by G. B. Airy. It accounts for the unexpectedly small deflection of a plumb bob near a mountain just as Pratt's theory does. Airy's is in better agreement with recent surface and subsurface information about rock composition. All the blocks in Airy's theory have the same density but "float" at different levels because they differ in height. The mountain block projects highest but also has the deepest root. Both theories can be called upon to explain certain areas. Pratt's seems best in explaining the differences between the ocean basins and continents and others. Apparently neither theory can stand alone to explain all crustal features. The specific gravities are recent, not known in 1855 when Airy proposed his theory.

tasy). If they were raised higher, they would simply sink to their proper floatation-supported height. Mountains higher than 20,000 feet such as the Himalayas or Hindu Kush are thought to be supported by the unusually thick mass of light, continental-type crust beneath them which was emplaced by the collision of India with Asia. They are probably floating on the combined crustal thickness of Asia and India.

Evidence for a state of flotational balance or isostasy can be read in the great masses of ice called **continental glaciers** which cover Greenland and Antarctica. The weight of the ice, which is more than 2 miles thick, has been compensated for by the down-warping of those landmasses. The Scandinavian peninsula was covered by such an ice sheet up until some tens of thousands of years ago, when the ice melted; since that time the land has been raised slowly by isostatic rebound as indicated by old beaches which, by now, have been raised more than 1500 feet above sea level. The peninsula is still rising slowly because isostatic balance has not yet been attained. It is estimated that the Baltic area will continue to rise over 500 feet more, but at the rate of a fraction of an inch per year. Crustal movement downward is taking place now in the Nevada-Utah-Arizona-California area surrounding Lake Mead. Here Boulder Dam backed up water to form a lake which placed additional weight on that part of the crust, pushing it deeper into the mantle.

We might view the earth as composed of large divisions, such as continents, oceans, continental-size glaciers, and the crust underlying them all in balance with one another and the denser mantle material beneath them on which they seem to float. In a real sense, the crust floats on the mantle, sinking deeper into the mantle where the crust is thick and floating higher where it is thin. The base of the crust beneath mountain ranges is deeper than beneath low plains.

As material is eroded from the continents and transported to the oceans, the isostatic balance is upset. In response, the continents must rise and the ocean floors sink in proportion to the mass which they respectively lose and gain through the processes of erosion and deposition. Rocks of the mantle are believed to flow slowly in response to local differences in temperature or pressure. Thus the underlying mantle material must flow away from or into the area to compensate for the sinking or rising crust (see fig. 5.4).

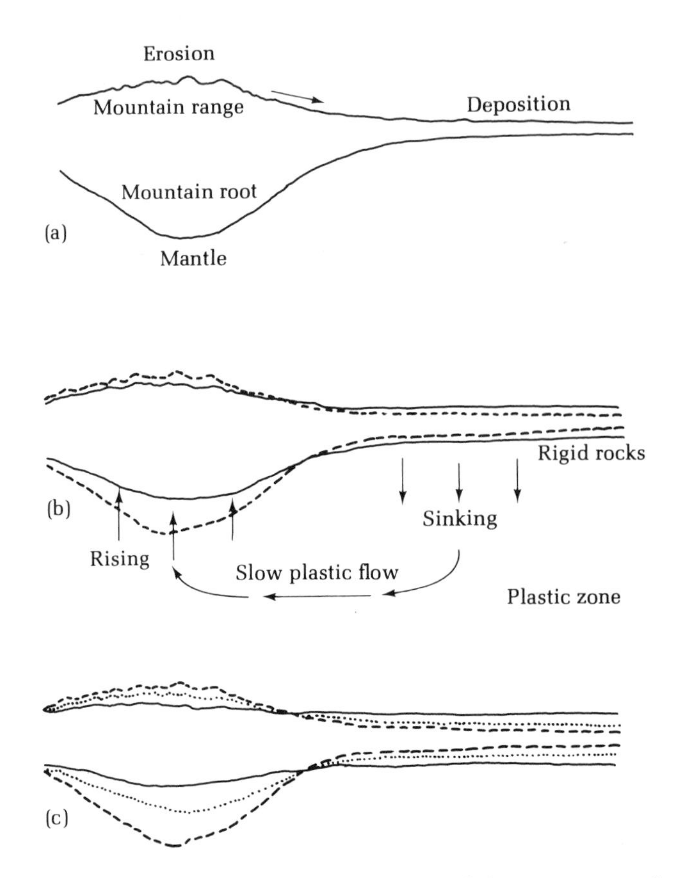

FIGURE 5.4 Three Stages in Isostatic Uplift. As material is eroded from the high land, that area becomes lightened and responds by isostatic uplift. Simultaneously, the low area receiving the sediment may be depressed by the additional mass. (Reprinted, by permission, from Robert J. Foster, *General Geology*. 2nd ed. Columbus: Charles E. Merrill Publishing Co., 1973. Courtesy of the author.)

Gravity Measurements

Information about the earth's gravity has been a useful tool to geologists not only in prospecting for mineral deposits but also in resolving certain aspects of the continental drift problem.

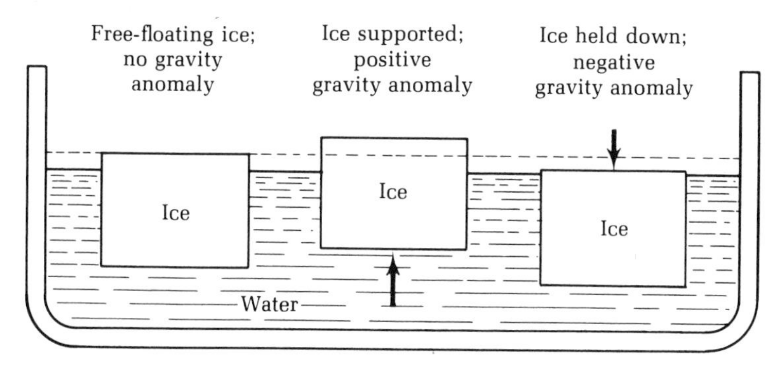

FIGURE 5.5 Gravity Anomalies. Ice block (1) floats freely and has normal gravity. Block (2) is pushed up, allowing more of the heavier water beneath it, thus producing a positive gravity anomaly. Block (3) is depressed with the lighter ice displacing the heavier water, thereby producing a negative gravity anomaly. (From Foster, *General Geology*, 2nd ed.)

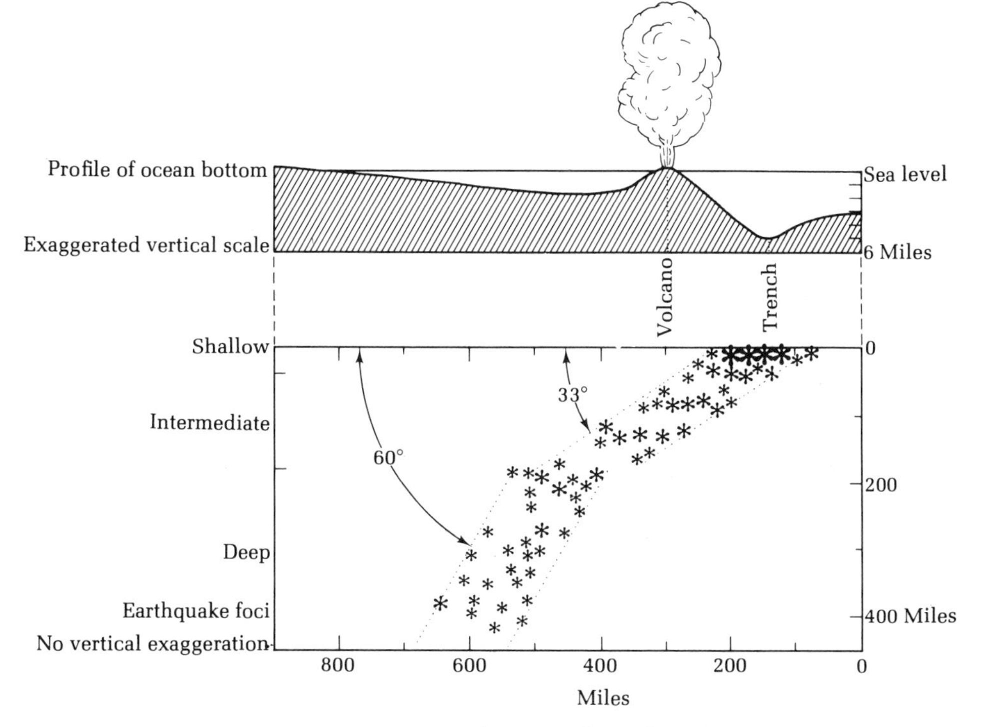

FIGURE 5.6 A Cross Section of a Trench with Volcanoes to Its Landward Side Shown in Exaggerated Vertical Scale. Below it is a cross section in natural scale through the earth to a depth of more than 400 miles showing the location of earthquake foci. The foci lie in a Benioff zone which probably defines a fault plane in the Kamchatka-Kurile area. (After Hugo Benioff, from Foster, *General Geology*, 2nd ed.)

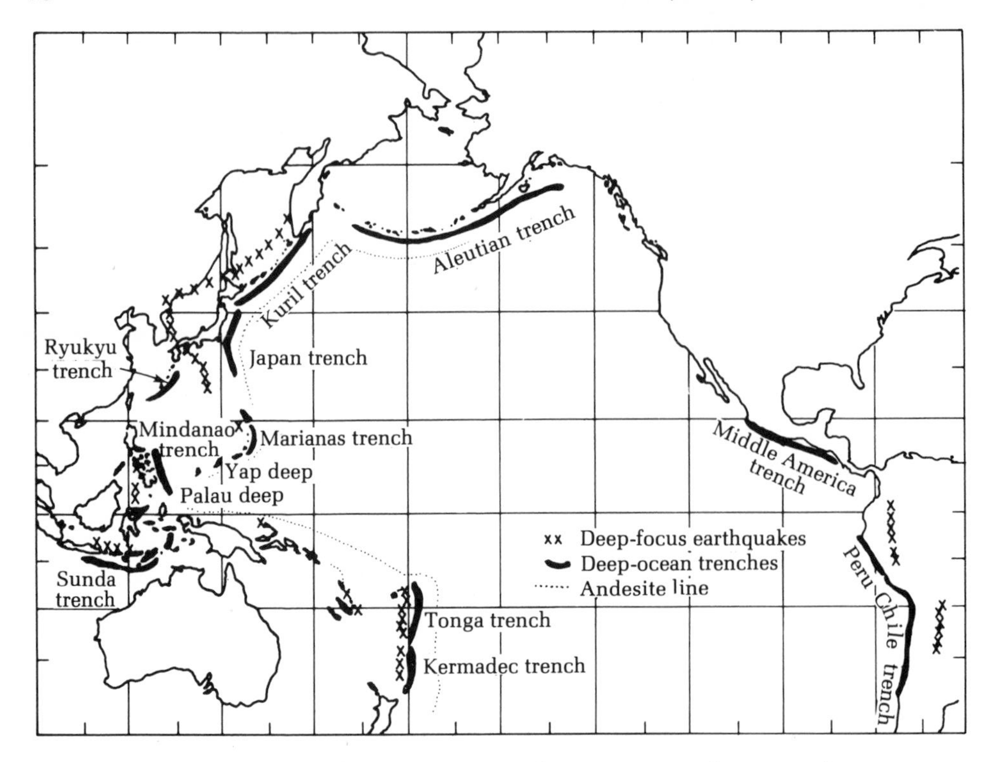

FIGURE 5.7 Pacific Basin Map Showing How the Trenches, Volcanic Island Arcs, and Deep-seated Earthquakes Are Related. The volcanoes on the oceanward side of the dotted line (andesite line) erupt basaltic lavas rich in iron and magnesium; the volcanoes on the continent side of the andesite line emit lavas with less iron and magnesium, but are richer in silica — these form the granitic rocks. (From Foster, *General Geology*, 2nd ed.)

Generally, the crust under the oceans is only about 5 miles thick and is composed of dense (2.9 density) basaltic rock. In contrast, the crust underlying the continents is very much thicker, about 15 to 20 miles on the average; it is largely made up of lighter (2.6 density) granitic rock.

When gravitational strength is measured over light rock such as a salt dome, it is found to be weaker than average for the earth as a whole; just the opposite is found over dense rock such as a metallic-ore body. Gravimeters are instruments used to measure gravitational intensity. Such intensity is expressed in milligals (one-thousandth of a gal). The earth's average gravitational strength at sea level is 980 gals. As a standard of reference for expressing gravity, the earth is considered a nearly perfect sphere with sea level as its surface. The gravity reading at any point on earth is then compared with the theoretical value found on a spherical earth. The difference between the actual and theoretical gravity values is called a gravity **anomaly.** The anomalies are almost always negative over land and positive above the ocean. (See fig. 5.5.) Additionally, the anomalies increase in mountain areas with increasing height above sea level.

Oceanographic surveys of the past fifty years revealed a worldwide distribution of ocean floor trenches or troughs which are roughly V-shaped, thousands of miles long and hundreds wide. Most are located along the margins of the continents, and almost all of them lie along the border of the Pacific Ocean. The trenches are bordered on the landward side by arc-shaped chains of active volcanoes (see fig. 5.6). The troughs form the deepest surface features on earth; they are gently curved and linear and are associated with frequent earthquakes and volcanic action. (See fig. 5.7.)

Many of the trenches have rocky bottoms and are bare of sediment which is found over most of the sea floor elsewhere. A rocky trench suggests that the crust must have subsided rapidly because the trenches are not far from the continental margins and thus even small amounts of sediment would tend to fill such depressed areas. The variation in the filling of the trenches on a worldwide basis has added further to speculation on their origin and histories. From 1923 through 1930, F. A. Vening-Meinesz performed the first gravity measurements at sea in a Dutch naval submarine in order to escape the disturbances of surface waves. After measuring the force of gravity over thousands of miles of deep-sea trenches in the East Indies and Philippines area, he discovered that the force of gravity was low over the deep-sea trenches and consequently concluded that the earth's crust was not in a state of flotational balance (isostasy) there. Instead, he believed that some force present in the trench areas was pulling the crust downward and holding it there against the force of gravity to form deep-sea trenches. It was as if a cork were being held down against its attempt to rise in water (negative gravity anomaly. See fig. 5.8).

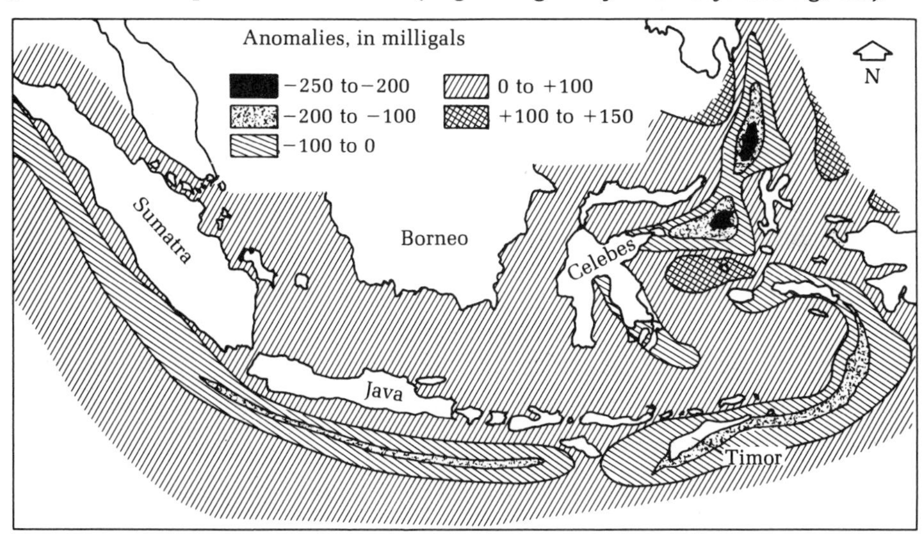

FIGURE 5.8 Gravity Anomalies and Oceanic Trenches. (After Vening-Meinesz, 1948, from Leigh W. Mintz, *Historical Geology: The Science of a Dynamic Earth.* Columbus: Charles E. Merrill Publishing Co., 1972. Courtesy of the author.)

Convection Currents

Vening-Meinesz had a ready-made theory in answer to his disturbing question of the mysteriously depressed crust of the trench areas before it was asked. The logical candidate was convection currents. Convection is the process which water undergoes when it is heated in a pan; the water near the heat expands, becomes lighter and rises; as it moves away from the heat, it cools, becomes heavy and sinks so that circulation in the fluid takes place and a convection cell or system is formed. (See fig. 5.9.)

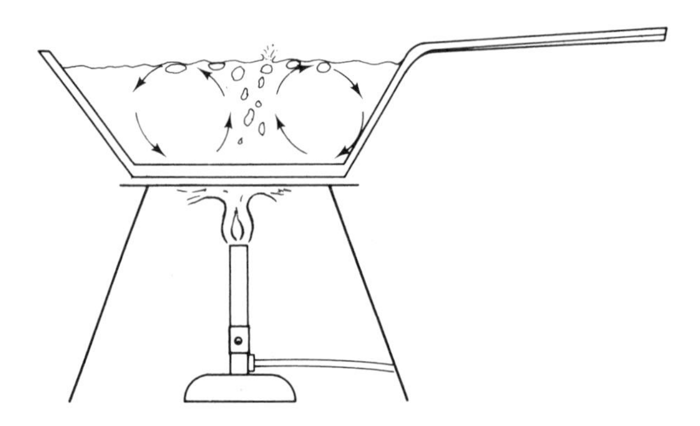

Water heated in a pan undergoes convection.

FIGURE 5.9 Convection. Water at bottom is heated, expands, becomes lighter, and rises. As it rises and moves from heat source, it cools, becomes heavy, and sinks, forming a pattern of circulation within the fluid called a *convection cell* or *convection system.*

The possibility of convection below the earth's crust was first suggested by William Hopkins in 1839. In 1881, Osmond Fisher used convection to explain mountain building, noting that the drag of such currents on the underside of the crust might produce local concentrations of heat and volcanoes. In 1906, Otto Ampferer of Austria called on convection currents to explain mountain building — he mistakenly believed that the earth's mantle was in a liquid state. Twenty years later, John Joly, the great Irish physicist, expanded Ampferer's ideas, but he too was unaware that the mantle was not liquid, as had been shown in 1899. In 1928, Arthur Holmes of Edinburgh University first suggested in modern terms the theory that the solid material of the mantle, in response to heat from the earth's interior, could undergo slow convection just as a liquid goes through rapid convection. Holmes' convection current hypothesis explained how the crust in the trenches was being held down against its natural tendency to rise. (See fig. 5.10.)

In a convection current system, material is carried down at one point, up at another, and sidewise elsewhere. Here not only did the "drifters" find a force to depress the crust under the deep-sea trenches and hold it there against buoyancy, but the same convection current could carry a continent across the earth's face! The drifters welcomed Holmes' hypothesis which is similar to the presently accepted one. In summarizing his discussion, Holmes noted a most important idea which later workers often modified and used in elaboration of his ideas: "During large-scale convection circulation the basaltic layer (dark, dense volcanic rock of the ocean floor) becomes a kind of endless traveling belt on top of which a continent can be carried along, until it comes to rest (relative

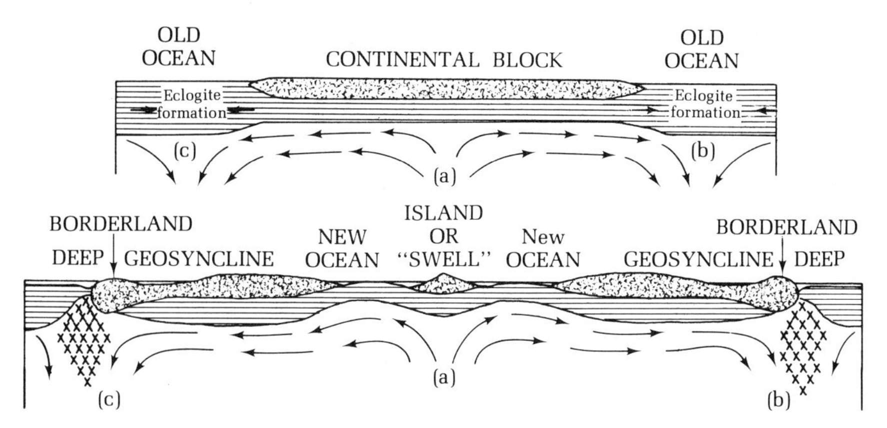

FIGURE 5.10 Arthur Holmes' Pioneering Concept of Convection Circulation in the Solid Mantle of the Earth. (a) A current rises vertically below the central part of a continent. (b) The current forks to produce two horizontal currents moving in opposite directions at the top of the mantle. (c) The opposed horizontal currents split the continent and carry off their respective parts. (d) Each of the horizontal currents collides with currents from other heat sources and is bent downward, pulling the continental margin into a depressed bulge. (e) When the convection current subsides, the continental crust which had been bulged downward will rise by simple flotational force (isostasy), much as a cork surfaces after being held below the water surface. (Reprinted, by permission, from Arthur Holmes, *Transactions of the Geological Society of Glasgow*, 1929.)

to the belt) when its advancing front reaches the place where the belt turns downwards and disappears into the earth." This theory was proposed by Holmes in 1928 but went almost unnoticed until the 1960s. Earlier theories which attempted to explain continental drift had the continents sailing across the top of the mantle; Holmes had the top of the mantle in motion with the continents being carried along on top of it — much as they would be carried along on top of a conveyor belt.

In 1962, Harry H. Hess of Princeton University wrote a now classic paper entitled "History of the Ocean Basins." He considered it an "essay in geopoetry," obviously to emphasize the then highly theoretical quality of the ideas presented. With Holmes' original hypothesis as a base, Hess departed to accentuate the plunging of the sea floor edges into the deep-sea trenches at the edges of the continents, noting that the older parts of the ocean floor would constantly be destroyed, thus accounting for the absence of truly ancient sediments and islands. He believed that the mantle is convecting at the rate of about 2/5 inch (1 cm) per year under the Mid-Atlantic Ridge, carrying the ocean floor sediments and volcanoes down into the trenches to be metamorphosed and welded onto the continents. Uniquely, Hess held that convective flow from the mantle "comes right through to the surface, and the oceanic crust is formed by hydration [water combining with rock material] of mantle material starting at a level 5 km [3 miles] below the sea floor." The peridotite of the upper mantle supposedly was converted to serpentine by hydration to form an ocean crustal layer. Perhaps most importantly, he summarized the geophysical and geologic data bearing on the origin of the ocean basins and provided the impetus for the younger generation of scientists who undertook the magnetic surveys of the sea floor which by the mid-1960s began to provide the truly conclusive evidence for sea-floor spreading.

Alternatives to Convection

The most widely accepted mechanism for moving the continents is still convection, very nearly in the form first proposed by Holmes and Vening-Meinesz, but the theory is weak in that convection does not take place along lines or offset broken lines such as those formed by the mid-oceanic ridges.

Another alternative theory is that the plunging part of the plate is cooler and more dense than the mantle into which it descends; therefore, it pulls the plate along behind it as it sinks deeper and is consumed. This idea appears difficult to apply to the South Atlantic mid-oceanic ridge because neither the African nor the American plate which spread from this ridge descends into the mantle at its opposite edges; thus the sinking and pulling theory does not hold up everywhere.

Still another alternative theory suggests that the sinking plate chills the adjacent mantle to produce the convection currents which drive the plates. This theory seems applicable to such areas as the nearly enclosed Sea of Japan which has a typical ocean floor covered by several miles of sediment. This suggests that the floor has been sinking for a long time. Perhaps a sinking current of chilled mantle material on the upper side of the plate depresses it to form such deep seas. Such small, enclosed seas are presently developing behind the island arcs, but older seas such as the Black and North seas and the Gulf of Mexico may be of like origin.

The movement of great plates of lithosphere and other lines of evidence all suggest a system of convection currents within the mantle of very broad dimensions. On the other hand, in a paper on the plastic layer of the earth's mantle D. L. Anderson has recently shown that convective flow appears to be limited to a zone only a few tens of miles thick. Recent experiments suggest that convection cells are not likely to develop in a space of such restricted shape. Therefore, attention has been refocused on another possible engine by which the continents have been propelled — that of hot spots or thermal centers. Tuzo Wilson first proposed that there are hot spots or rising plumes which arise from deep within the mantle and remain fixed for 100 million years or more. Recently W. J. Morgan has expanded on this idea, theorizing that these rising plumes or hot spots raise the lithosphere above them into domes about 2 or 3 miles high. Such upwellings are continually erupting lava on the ocean floor to form volcanoes. Perhaps the Atlantic mid-oceanic-ridge-rift zone is a crack which formed, connecting a series of such upwelling hot spots which may exist under the islands of Iceland, the Azores, St. Peter and St. Paul Rocks, Ascension, Tristan de Cunha, Bouvet, and Jan Mayen. It may be that the American, African, and Eurasian plates are merely sliding down the sides of the domes in both directions to be consumed in the trenches. A. L. Hales and W. R. Jacoby have indicated that gravity alone may be simply moving the plates downhill — this merely requires that the underlying asthenosphere be in a semimolten or plastic state with no permanent strength. The simplicity of the idea is certainly attractive, but convection theory has not yet been discarded by a long shot.

Heat from the Earth

Research during the past two decades on the question of heat arising from the earth's interior has led to discoveries which also support the idea of convection currents.

The rate of temperature increase with depth is called the **geothermal gradient.** The amount of heat energy flowing through a unit of area in a unit of time is called **heat flow;** this energy is released in the earth by heat sources. Heat is generated by the decay of radioactive elements of which uranium, thorium, and potassium are most important. Even the least radiogenic, near-surface rocks contain so much radioactivity that it appears that if the whole earth below the crust were similarly composed, the earth would be molten. A solid mantle, therefore, indicates that most radioactivity is concentrated in the crust. The average granite rock produces so much heat that a world-encircling, granite layer only 8 miles thick could supply all the heat which the earth presently radiates into space. The continents are largely underlain by granitic rock which is a much greater heat producer than the basaltic rock which underlies the ocean basins. In spite of this, the heat flow from the ocean basins is higher than from the ancient granite shield or nuclei areas of the continents and equal to the heat flow from the Paleozoic mountain chains. A large part of the heat flow from the continents must be derived from the continental crust. The basalts of the ocean floor are poor heat generators and are therefore not responsible for the heat flow there; but surprisingly the oceanic flow is comparable to that of the continents. Thus, most of the heat coming through the rocks of the ocean floor must be generated in the mantle. This suggests that the mantle under the oceans is very different from that beneath the continents. Heat flow is measured in heat flow units (HFU) in which 1 HFU equals 10^{-6} cal/cm^2 sec. Heat flow measurements have been made over about one-half of the earth's surface; most measurements are between one and two heat flow units. The most frequent value was 1.3 HFU with most readings below 0.6 HFU recorded in the oceans. The values above 2–2.5 HFU usually represent areas of recent hot springs and volcanic activity.

Low and generally uniform heat flow is found in the deep-ocean trenches, and continental areas which have been geologically stable for hundreds of millions of years. The flow is very high in seas recently created by the fracturing and separation of continents which the Gulf of Aden-Red Sea and Gulf of California are thought to represent. The heat flow is also very high along the mid-oceanic-ridge spreading centers. The heat-flow pattern across Japan and its volcanic island arc and nearby trench are shown in fig. 5.11. The variation is great with flow greater than 2 HFU near the string of active volcanoes and to the south down the Izu-Bonin arc. The flow falls to about 1 HFU along the eastern edge of the islands, and to the east the lowest values are found in the trench. It is now thought that the low values of the trench are caused by a descending portion of a convection cell within the mantle pulling down a piece of cool oceanic lithosphere which dives at about a 45° angle to produce the narrow region of earthquakes called a **Benioff zone.**

Heat flow measurements made along the East Pacific Rise and the great Mid-Atlantic Ridge show values two to eight times normal. The flow is great at the ridge crest but decreases away from the center, falling to below normal along the flanks of the ridge. It is difficult to explain why the ocean floors and continents show nearly equal heat flow when the granitic continents are known to have greater heat-producing capacity. It was thus suggested that heat is generated in the mantle beneath the oceans and carried to the surface by con-

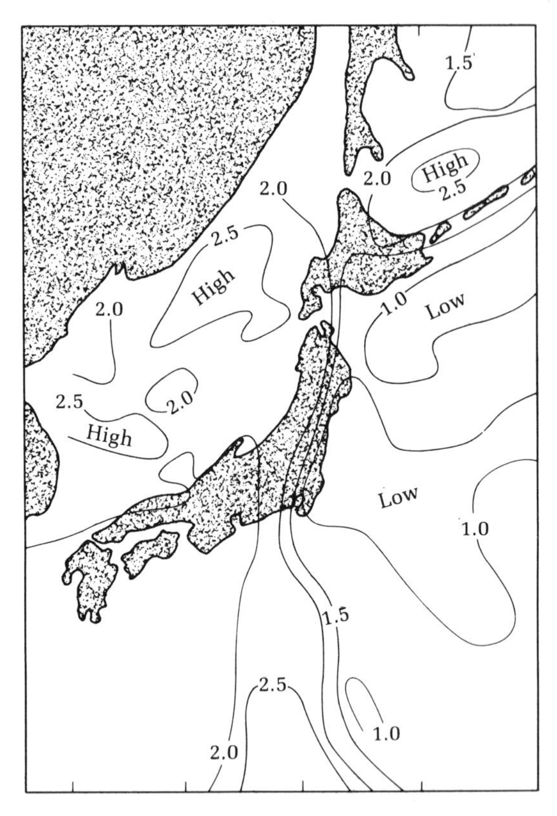

FIGURE 5.11 The Heat-flow Pattern across Japan and Its Volcanic Island Arc and Nearby Trench. The low heat flow is due to a piece of crust diving along the earthquake fault zone (Benioff zone). The string of volcanoes along the spine of the Japanese Islands is associated with high heat flow and great hot-spring activity there. (Reprinted, by permission, from Seiya Uyeda and Victor Vacquier, *Geophysical Monograph 12,* The Crust and Upper Mantle of the Pacific Area, Figure 2, p. 352, 1968, copyright by American Geophysical Union.)

vection within the mantle. The greatly heightened heat flow beneath the crests of mid-oceanic ridges is now thought to be supported by mantle convection. We might expect that the hot, rising convection column be located beneath the ocean ridges while the sinking, cool part of the convection cell should turn down, dragging the lithosphere to produce a deep-ocean trench. (See fig. 5.12.) Heat-flow studies support the convection idea.

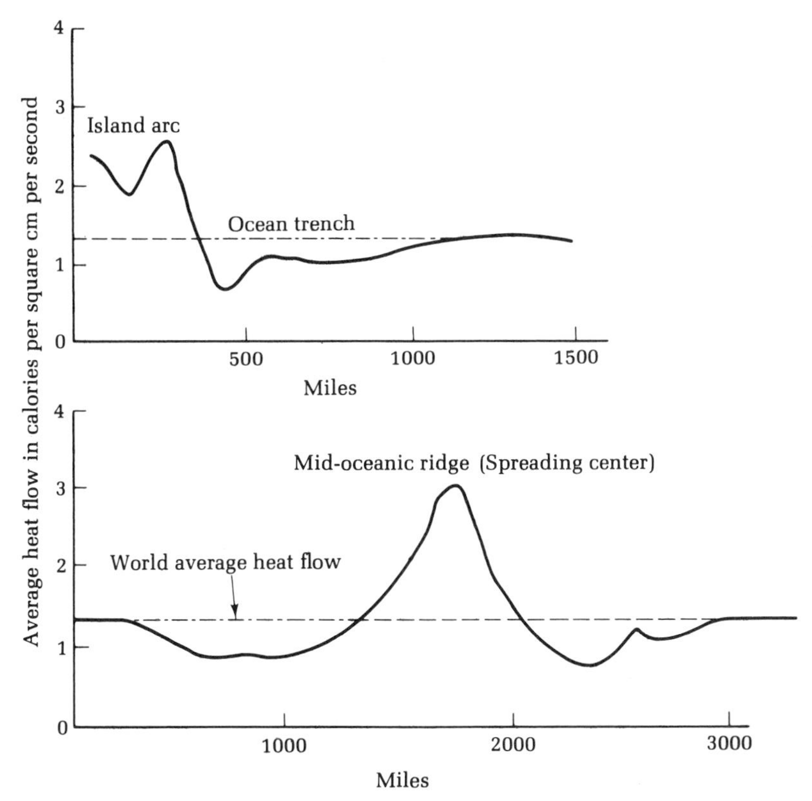

FIGURE 5.12 The Heat Flow from Mid-oceanic Ridges and Deep-sea Trenches Showing Much Greater Flow over the Ridges and Considerably below Average Flow over the Trenches. (Reprinted, by permission, from Don and Maureen Tarling, "Continental Drift." London: G. Bell & Sons, Ltd., 1971.)

The volcanoes which border the deep-ocean trenches apparently are fed by lava coming up through fault zones that typically extend from the inner side of the ocean trenches and dip toward the neighboring continent down to a depth of about 400 miles.

During the period in which Vening-Meinesz performed his gravity measurements, exciting work was undertaken by seismologists in measuring and interpreting shock waves produced by earthquakes originating in the fault zones which lie along the landward side of the ocean trenches under the volcanic island arcs. The distribution of earthquakes follows a pattern (see fig. 5.7).

In front of the arc, between the islands and the ocean trench, the earthquakes have a normal, shallow point of origin (focus). However, the earthquake foci become increasingly deep as we move further behind the arc, as shown in fig. 5.6. If all the foci are laid out in a cross section, they will lie approximately in a plane which dips toward the continent side of the arc at an angle of about 45°. This inclined, narrow region of earthquake production is common to all trenches. They are termed **Benioff zones** after Hugo Benioff who first called attention to them. Studies of earthquake shock waves produced in the zone indicate that the entire Benioff zone sinks relative to its surroundings. Benioff zones lie inclined below most of the world's volcanoes which are situated in the island arcs and continental margins. Apparently the whole arc structure extends at least as deep as the 600 kilometers (about 400 miles), which is the depth of some of the earthquake foci. The arc structures with their associated volcanoes and major fault zones are a major region of crustal instability. As we move from one island-arc area to another, minor differences are present, but the same general pattern prevails. Arc-like features are found also in mountain areas such as those along the west coast of South America where the ocean deeps lie offshore and the earthquake pattern is similar to that of the island arcs. The shallow quakes occur between the mountains and the deeps, and the deep-seated earthquakes are produced behind the arcs with their depth increasing with distance from the arc.

The two major earthquake zones on earth are the circum-Pacific and the Alpides, which include the Alps, Himalayas, and related mountains of Asia; all folded mostly during the past 60 million years. It was early noted that such earthquake zones lie along strings of island arcs and mountain ranges. Most of the volcanoes and earthquake zones of the world are associated with them. Apparently the arcs, volcanoes, and quakes result from the same basic forces. The arcs, with their associated ocean trenches, mountains, and volcanoes, are strung out over thousands of miles; earthquakes take place along their entire length. This suggests that the forces which produce the elongate arcs must operate at right angles to them. Large segments of the crust drifting across the earth would produce such features along their leading edges. It appears logical to assume that the oceanic crust is being carried downward in the zone of these earthquakes. The shallow earthquakes probably result from the cracking of the brittle oceanic crust. As the crust is carried deeper into the mantle, the greater pressure there turns the brittle rocks plastic; therefore, the deeper quakes must result instead from chemical and physical changes in the crustal rock as it comes under great temperature and pressure with depth. Below the depth of about 500 miles, the crustal rock becomes mantle rock and no earthquakes occur.

If we had started out very early with the premise that convection currents in the mantle did exist and then asked ourselves what features such currents should be expected to produce, we might have come up with a model very close to what we have just described. The crust is depressed to form ocean trenches and held down against the force of gravity like a submerged cork by the cool, descending portion of a convection cell. The heat flow is low over that part of the convection cell. The crust is raised to form oceanic ridges and rises above the hot ascending part of a convection cell. The heat flow is several times greater there than elsewhere on earth. We shall soon examine additional, more recent evidence from the continents and ocean floors which supports the concept of convection currents in the earth's mantle.

6

Fossil Magnetism in Rocks

The Earth as a Magnet

The most convincing evidence about continental drift is contained within certain rocks of the continents and ocean floors. Locked in the rocks are fossil compasses which actually point the way to the former positions of the continents; it is this evidence, more than any other, that has turned the tide of scientific opinion in favor of continental drift.

The common compass is simply a light bar magnet, balanced on a pivot point so that it may swing freely and align itself with the earth's magnetic field. One end of such a magnetic compass is attracted to the earth's north magnetic pole, and the other seeks the south magnetic pole. The earth itself behaves like a huge magnet. Although the compass needle was used for hundreds of years before, it was not until 1600 that Sir William Gilbert noted that the earth's magnetic poles were near to but did not coincide with the rotational or geographic poles. Why the earth has a two-pole or dipolar magnetic field is not really known, but a theory advanced in 1954 and 1955 by E. C. Bullard and W. M. Elsasser, each working independently, has gained acceptance. They think that the motion of the earth's rotation drags convection currents in the fluid outer core of the earth in the direction of rotation. The movement of the assumed molten iron of the outer core thus generates a magnetic field as if it were a self-exciting dynamo.

If the currents within the liquid outer core moved symmetrically about the earth's rotational axis, then we should expect the magnetic and rotational poles to coincide, but they do not. It appears in theory that the main two-pole magnetic field is upset slightly by eddies in the flow within the core. Therefore, the actual magnetic poles are more than 500 miles from the geographic or rotational poles. The north magnetic pole is about 15° from the geographic North Pole, and the south magnetic pole about 20° from the geographic South Pole. The earth's magnetic field varies slightly from year to year and place to place.

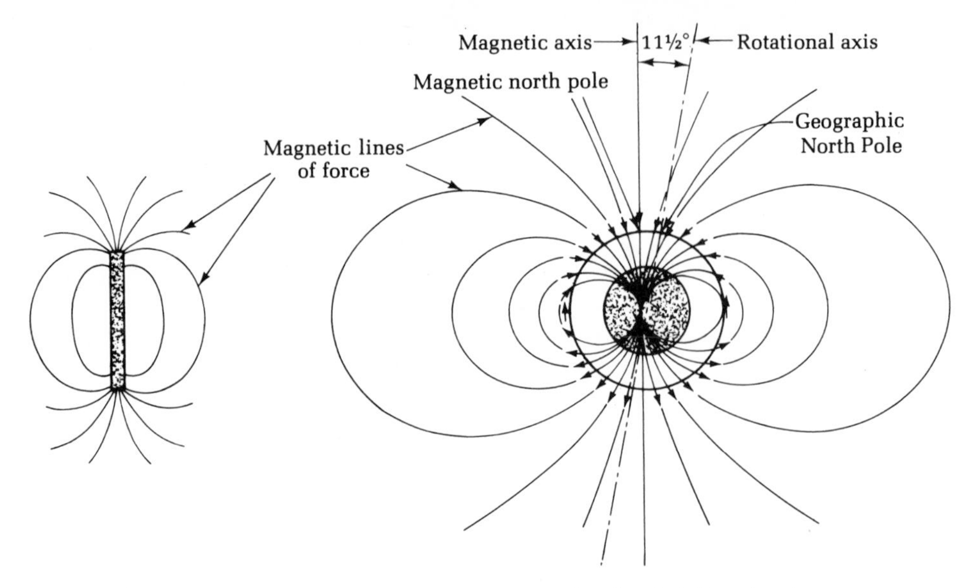

FIGURE 6.1 The Earth's Magnetic Field. The field is similar to a bar magnet; the lines of magnetic force are parallel to the earth's surface only at the magnetic equator and dip more steeply as the poles are approached (see fig. 6.2). The magnetic and geographic (rotational) poles do not coincide. (Reprinted, by permission, from Robert J. Foster, *General Geology*. 2nd ed. Columbus: Charles E. Merrill Publishing Co., 1973. Courtesy of the author.)

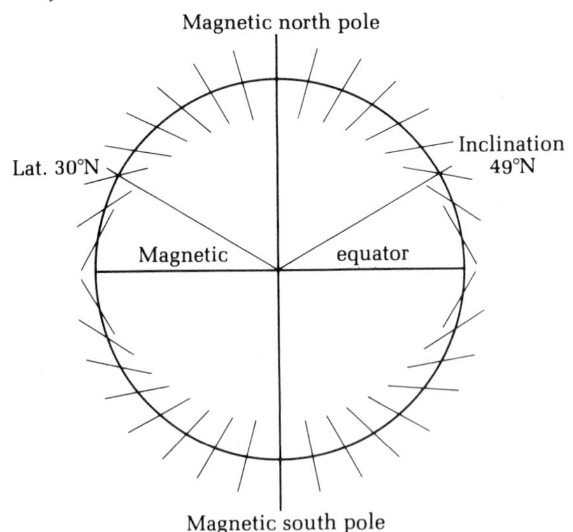

FIGURE 6.2 The Relationship between Magnetic Inclination and Latitude. The series of lines cutting the earth's perimeter show the actual angles at which a free-floating compass needle will dip below the horizontal plane. The needle will be in a horizontal position only at the magnetic equator. The compass inclination is related to latitude; e.g., at a latitude of 30° the compass needle dips 49°.

If iron filings are sprinkled on a sheet of glass overlying a bar magnet, we may visualize, in a greatly simplified way, the magnetic force field surrounding the earth (see fig. 6.1). Note that if we think of the earth as a sphere with a bar magnet within it, the lines of magnetic force parallel the earth near the magnetic equator only and dip more steeply as the poles are approached.

Thus, if a compass needle is balanced at its center of gravity, it will point to the magnetic north pole but will not come to rest in a horizontal position. The north-seeking end dips downward in the Northern Hemisphere and the south-seeking end dips downward in the Southern Hemisphere. The angle at which the needle dips varies with latitude. It is horizontal only at the magnetic equator and dips more steeply as the magnetic pole is approached (fig. 6.2.) The needle would dip vertically downward at the pole. A compass reading, therefore, has two parts — a horizontal component and a vertical one.

Magnetizing Rocks

The force of the earth's magnetic field is not very strong, but it is sufficient to magnetize certain kinds of both igneous and sedimentary rocks, which contain iron and nickel minerals. A ferromagnetic substance differs from others in that it can be magnetized in a magnetic field and retain its magnetization even when removed from that field. Ferromagnetic substances lose their magnetism if heated above the Curie point, which is 770°C for iron and 360°C for nickel. When molten rock or lava (about 1000°C) cools, its magnetic minerals crystallize; as the temperature falls below the Curie point, the crystals become magnetized parallel to the local magnetic field present at that time (fig. 6.3). In rocks such as basaltic lava flows, the crystals of ferromagnetic minerals such as magnetite and limenite become magnetized and show stability in retaining that magnetism.

Some minerals are magnetized exactly opposite to others, but they are rare and now recognizable. The ferromagnetism of such rocks is stable and does not change later even if the rocks are subjected to changes of the earth's magnetic field. In order to destroy or alter such a magnetic record, the rock must be heated above the Curie point. Such reheating of rock would lead to remagnetization and alignment with the new magnetic field of that later time, but the reheating would be recognizable.

Sedimentary rocks may also be magnetized and serve as fossil compasses. Sediment derived from preexisting rocks through erosion is transported by rivers and streams. It is deposited in horizontal layers in the lower reaches of river valleys, in lakes, and in greatest abundance in shallow water on the continental shelf and slope which border the continental edges. Some of the sedimentary particles contain iron such as magnetite grains. When deposited in a loose, wet, unconsolidated mass, these grains orient themselves parallel to the earth's magnetic field once again. Often iron-rich decay products of the erosion process, such as the mineral hematite, coat other mineral and rock grains and cement them together. The hematite, in a chemical rather than granular form, gives rise to chemical magnetization often found in red sandstone and shale beds.

Measuring the fossil or remnant magnetism in rocks is done with special, very sensitive galvanometers. Measurements are made on rock slices cut from cores which are drilled on the bedrock in place and carefully marked as to

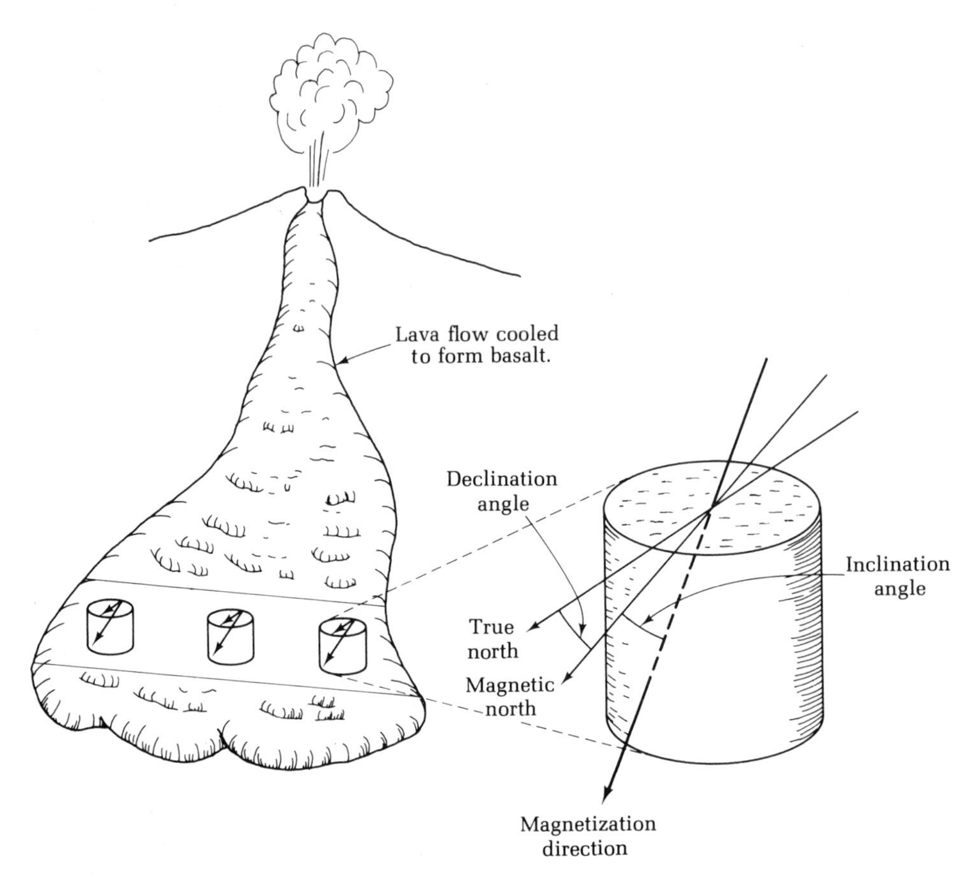

FIGURE 6.3 Rock Cores Drilled from Magnetized Volcanic Formations. The direction of magnetism is the actual direction in which the magnetic minerals within the rock were aligned at the time that they cooled and passed through the Curie point. The magnetization direction is expressed as the angle between true and magnetic north, and the inclination angle marks dip below the horizontal. (After Cox, Dalrymple, and Doell, 1967.)

original orientation. In order to insure reliability, specimens are usually taken from a number of different rock beds covering at least several thousand years.

All interpretations based on paleomagnetism accept the basic assumptions that the earth's magnetic field has always been essentially like that of the present one, e.g., a bar-magnet-like field, and that the magnetic poles have always been within about 15° of the geographical or rotational poles. In spite of the fact that the magnetic pole position has changed over the last few thousand years, its **average** position coincides with the rotational pole; this is thought to be true for earlier geologic time.

Rocks as Compasses

Data on the paleomagnetism of rocks from all over the world are accumulating which now permit the writing of a fairly detailed history of changes in the strength and direction of the earth's magnetic field through much of geologic time.

The latitude in which a rock became magnetized can be determined from the angle of dip below the horizontal shown by its fossil magnetic direction (see figs. 6.2 and 6.3). The direction of the ancient pole is also indicated. Thus, if the distance and direction of the pole from a specific locality are known, we may locate the pole on the earth at the time when the given rock sample became magnetized. Paleomagnetic information tells us the direction and distance to the ancient magnetic pole but **not** the longitude. Thus, the continent may be anywhere on the globe at the indicated latitude. A new method for determining longitude has been used recently by Dietz and Holden in a reassembly of the continents discussed in the plate tectonics chapter.

Movements of the Magnetic Poles

If magnetized rock samples are taken from a sequence of younger to older beds on a single continent, at each successive age the compasses "frozen" in the rock point to a different magnetic pole position. A line drawn connecting the indicated pole positions for the successive periods forms a "polar-wandering curve." Polar-wandering curves drawn from rock samples taken in North America and Europe are quite similar in shape but do not coincide for the time interval before about 75 million years ago (see fig. 6.4).The curve drawn from the European rocks plotted on a map lies in a position to the east of the North American one. In order to make the two curves coincide, the positions of the continents must be changed; they must be brought together, thus closing the Atlantic Ocean. To say that the curves **coincide** means that the magnetic particles in the rocks of the same age on the two continents point to the same magnetic pole.

In 1953 a London University group, which included Nobel laureate P. M. S. Blackett, found that the English Triassic rocks (225–180 million years old) were not lined up with the earth's present magnetic field. They also measured magnetism in the basalt rocks of India's Deccan Plateau which formed during the last 180 million years. They found it necessary to rotate England 30° clockwise and move it south and also move India to the Southern Hemisphere in order to satisfy their fossil magnetism data.

At approximately the same time, S. K. Runcorn, heading another group of workers, measured ancient magnetism in rocks all over Europe. Runcorn and paleomagnetists from Newcastle University drew paths for the poles indicated by evidence from the various continents and found that they did not coincide. The polar-wandering paths of Runcorn and others do not necessarily mean that the poles have moved; such polar-wandering curves may alternatively indicate that the continents have been moving relative to a fixed pole.

Inasmuch as the polar curves are different for each continent, we assume that the continents must have moved relative to each other. (See fig. 6.5.) The term **polar wandering** may mean that the entire earth body turns over relative to the axis of rotation. It may also mean that the outer shell of the earth slides over the interior. We shall learn in later chapters that the evidence strongly supports continental drift but not necessarily polar wandering. The present view has been stated by D. van Hilten: "According to our present knowledge, both processes may have played their part in the geological history of the earth. We are,

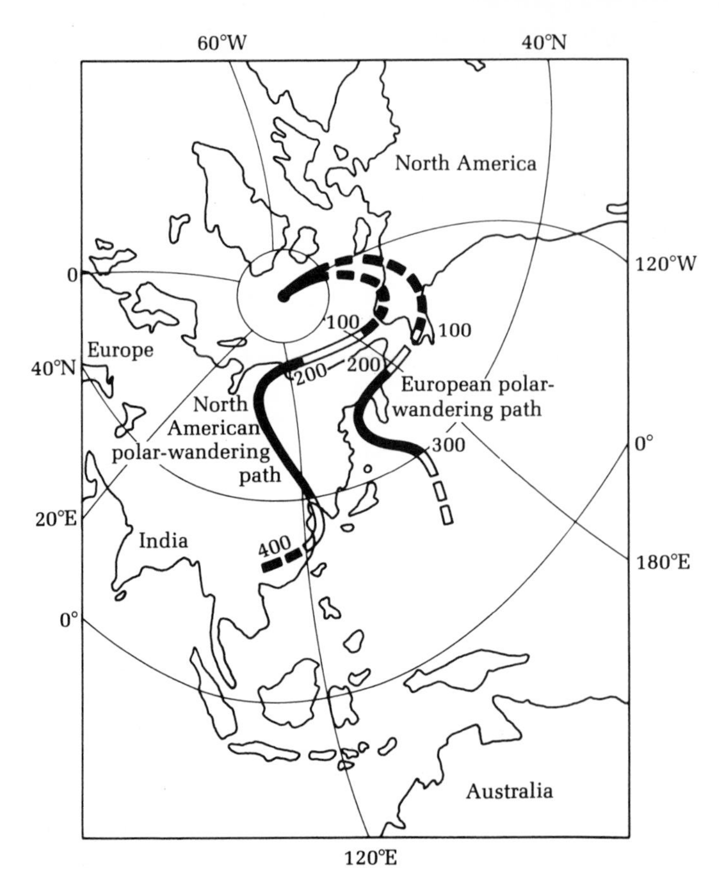

FIGURE 6.4 Polar-wandering Paths for North America and Europe. A sequence of magnetized rock layers of successive ages from each continent establishes two different paths. If the continents are brought together, closing the Atlantic Ocean, the curves then coincide. For the period between 100 and 300 million years ago, the distance between the two paths is equal to the width of the modern Atlantic Ocean. The numbers (millions of yr ago) are located at ancient pole positions along the paths (solid line from data, broken line inferred). (Reprinted, by permission, from Don and Maureen Tarling, "Continental Drift." London: G. Bell & Sons, Ltd., 1971.)

however, completely at a loss about their relations. Are both processes occurring contemporaneously or in some succession, and is there any causal relation at all?" P. M. S. Blackett suggests that the term **continental drift** should, for the present, include polar wandering as a component until science is able to recognize the effects of each separately. It is interesting to note that when the continents are moved about in order to satisfy the fossil magnetism evidence, such relocation is perfectly in keeping with the evidence of ancient climates.

There is evidence that North Africa was covered by continental glaciers about 500 million years ago. This idea agrees with magnetized rocks of the same age from Africa which point to the Sahara Desert as the location of the pole at that time! About 200 million years later, South Africa appears to have been the site of a polar ice sheet which is also in agreement with the fossil magnetism evidence which indicates that the pole had wandered to South Africa by that time. We may cite many examples to show that there is close agreement between the paleoclimatic and paleomagnetic evidence.

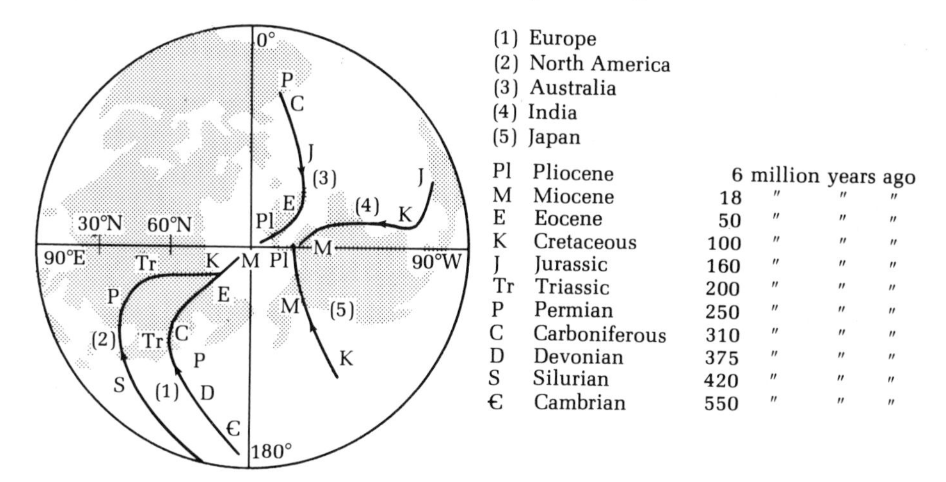

FIGURE 6.5 Polar-wandering Paths Shown for Five Continents from Cambrian through Pliocene Time (about 550 Million Yr). Each numbered line represents the path along which the magnetic pole moved relative to the present position of each continent; the letters show the position of the pole on the path during successive geologic periods. Since the polar curves are different for each continent, it indicates that the continents have moved relative to each other. (From Allan Cox and R. R. Doell, Courtesy of the Geological Society of America, from *Bulletin* 71, 1960.)

Magnetic Polar Reversals

The first rocks showing reversed magnetism were discovered in 1906, giving rise to the question of whether the earth's magnetic field had been reversed in the past in order to produce such reversely magnetized rocks. Mercanton first argued in 1926 that the earth's magnetic field had reversed itself, and he encouraged workers on other continents to seek evidence for reversals. He clearly stated at that time that fossil magnetic information might test the hypothesis of continental drift and polar wandering. Almost all magnetic evidence now shows that the earth's magnetic field has indeed reversed itself repeatedly throughout geologic time. The North Pole has become the southern one, and the South Pole has become the northern one. In the 1950s it was further demonstrated that all magnetized rocks of the same age show the same polarity. The development of the potassium-argon method of rock dating has allowed critical study of geologically young rocks. These studies show that there have been long intervals during which the magnetic field was the same as now; these periods alternate with equally long epochs during which the field was reversed. The major polarity epochs have been named for major contributing geophysicists (see fig. 6.6).

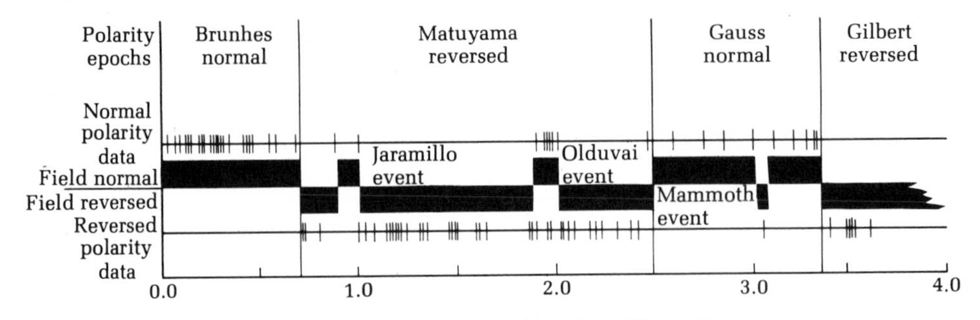

FIGURE 6.6 Times of Normal and Reversed Magnetic Polarity on the Earth during the Past Four Million Years. (After Allan Cox and G. B. Dalrymple, 1967.)

The Bruhnes normal, the youngest polarity epoch, extends from about .7 million years ago to the present; before the Bruhnes normal, the Matuyama reversed epoch lasted from 2.5 million years to .7 million years ago. Note that the Matuyama has two brief reversals to normal polarity within it; these are the Jaramillo and Olduvai events. Preceding the Matuyama was the Gauss normal epoch from 3.36 to 2.5 million years ago, which was punctuated by the brief

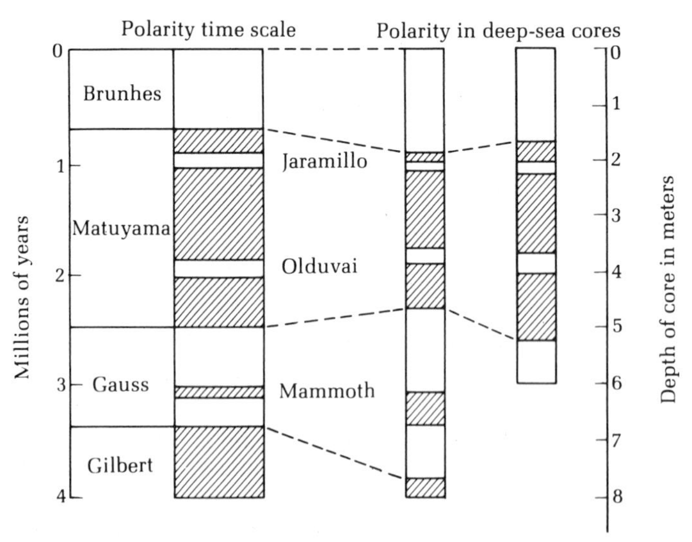

FIGURE 6.7 Magnetized Sediments in Deep-sea Cores. The sediments show normal (dark) and reversed (light) layers that correlate with the polarity time scale determined from volcanic rocks on land. The cores shown are from Antarctic waters. (After Cox, Dalrymple, and Doell, 1967.)

Mammoth reversal event. The Gilbert reversed epoch extends back from 3.36 million years to an indefinite point before about 4.0 million years ago. We cannot be more specific because dating methods do not permit us to distinguish small time intervals that long ago.

C. G. A. Harriman and B. M. Funnell of Scripps Institution of Oceanography, and N. D. Opdyke and D. E. Hayes and workers at the Lamont-Doherty Geological Observatory have shown reversals in sediments taken with deep-sea coring devices dating back to 3.6 million years ago. The polarity reversal pattern of the sediments in the oceanic sediment cores is very similar to the polarity time scale worked out by Allan Cox and his colleagues from continental volcanic rocks. The magnetic time scale derived from land rocks can then be applied to ocean-floor-dating problems. Such land-sea correlations provide a key for determining sedimentation rates and drawing comparisons among oceanic sediment cores from widely separated localities (see fig. 6.7).

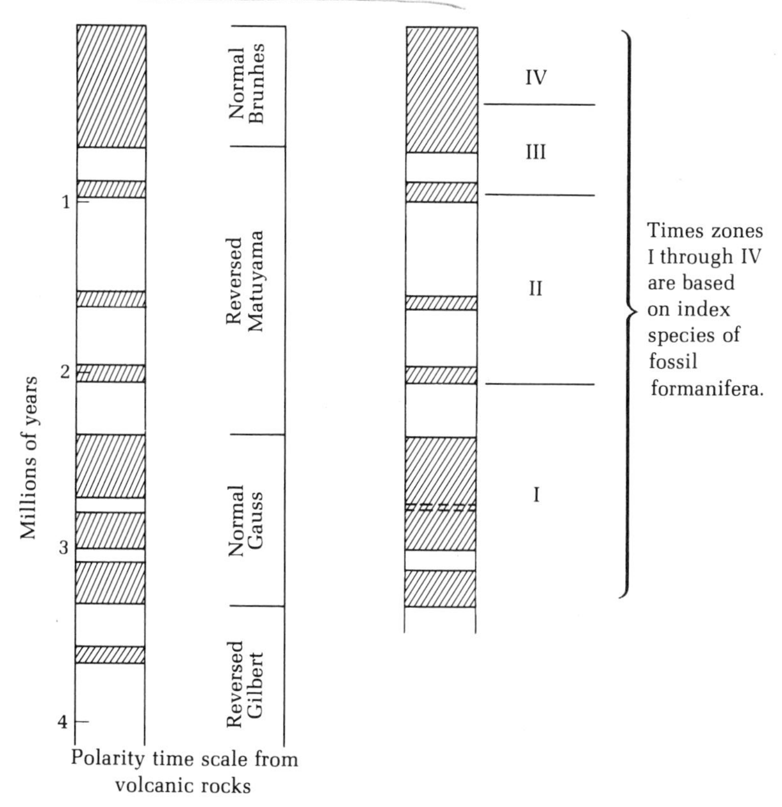

FIGURE 6.8 Correlation of Magnetic Polarity between Volcanic and Sedimentary Rocks. The same magnetic polarity occurs in volcanic rocks on land dated by the potassium-argon method and sedimentary rocks dated by fossils. Index species of fossil foraminifera facilitate the recognition of fossil-time zones. Since igneous and sedimentary rocks have very different origins and methods of dating, the close match between their ages and polarity strongly supports the concept of polar reversals. (After Tarling, 1971.)

The polarity and ages of the volcanic rocks dated by the potassium-argon method appear to be in close agreement with magnetized sedimentary rocks which were dated by their fossil foraminifera content (see fig. 6.8). In this independent and separate method of dating by fossils, we gain a high confidence level as we find that the index fossil species of foraminifera allow us to identify time zones in the sedimentary layers during which the beds were magnetized in the same way as contemporaneous igneous rocks forming elsewhere on earth. Since igneous and sedimentary rocks have very different origins and methods of dating, the close match between their ages and polarities strongly supports polar reversals.

The time scale for reversals of the earth's magnetic field has been worked out by Allan Cox and G. B. Dalrymple and others who believe that the average time involved in any one of the many reversals that they have recorded is somewhere between 1600 and 21,000 years, with 4600 years given as a best estimate. Their work on the geologically young rocks of the continents during the 1950s really provided the necessary base on which later investigators stood in examination of magnetized rocks of the sea floor.

It now appears that the earth's magnetic field has been diminishing during the past 140 years and will fall to zero in about 2000 years. It is thought that such a decrease is part of the reversal process which we may now be undergoing and that after collapse, the field will rebuild itself with opposite polarization. The possible impact of such reversals on living things is discussed in our chapter on fossil life.

Polarity reversals seem to explain the striped pattern of magnetic variations found on the sea floor in many areas of the world. These magnetically banded rocks which extend for thousands of miles in some places appear to be symmetrically distributed around the crests of the mid-oceanic ridges. This most recent line of important evidence is discussed in our next chapter.

7

Plate Tectonics

The Plate Boundaries and Earthquakes

The term **tectonic** refers to the deformation of the earth's crust and the rock structures and surface features resulting from such movement. Plate tectonics has grown out of studies on continental drift and may be thought of as its offspring. The theory of plate tectonics holds that the outer shell of the earth, or **lithosphere,** is composed of about a dozen large and several smaller rigid slabs or plates which are fitted against each other along boundaries which are of three kinds (fig. 7.1): (1) oceanic ridges along which plates move apart and new crust is generated; (2) transform faults which allow plates to slide past each other; and (3) subduction zones at which plates collide with the result that the leading edge of one plate dives under the leading edge of the other and crust is destroyed. The most complex meeting place between the plates occurs at a triple junction where three plates meet. Such junctions may involve various combinations of the three types of plate boundaries.

Most earthquakes occur in belts which lie along the tectonic-plate boundaries. Although there are three kinds of plate boundaries, the earthquake zones seem to be divisible into four classes: (1) the zone which lies along the mid-oceanic-ridge spreading centers at which plates depart with the production of shallow earthquakes; (2) the transform fault zone, exemplified by the San Andreas fault along which shallow earthquakes occur as plates move past each other; (3) the subduction zones or deep elongate troughs with their volcanic island arcs, marked by deep, intermediate, and shallow quakes. The zone of quakes here dips below the trench toward the continent and is associated with volcanic chains; (4) the quake region which lies in the area of major compressed mountain ranges such as the great Himalayas and Hindu Kush. Such a region extending from the Mediterranean almost to China generally shows shallow quakes with some intermediate and still fewer deep ones.

Very recent study of the earthquake belts of the world suggests that the mid-oceanic ridges are under tension. The second type of plate boundary, such

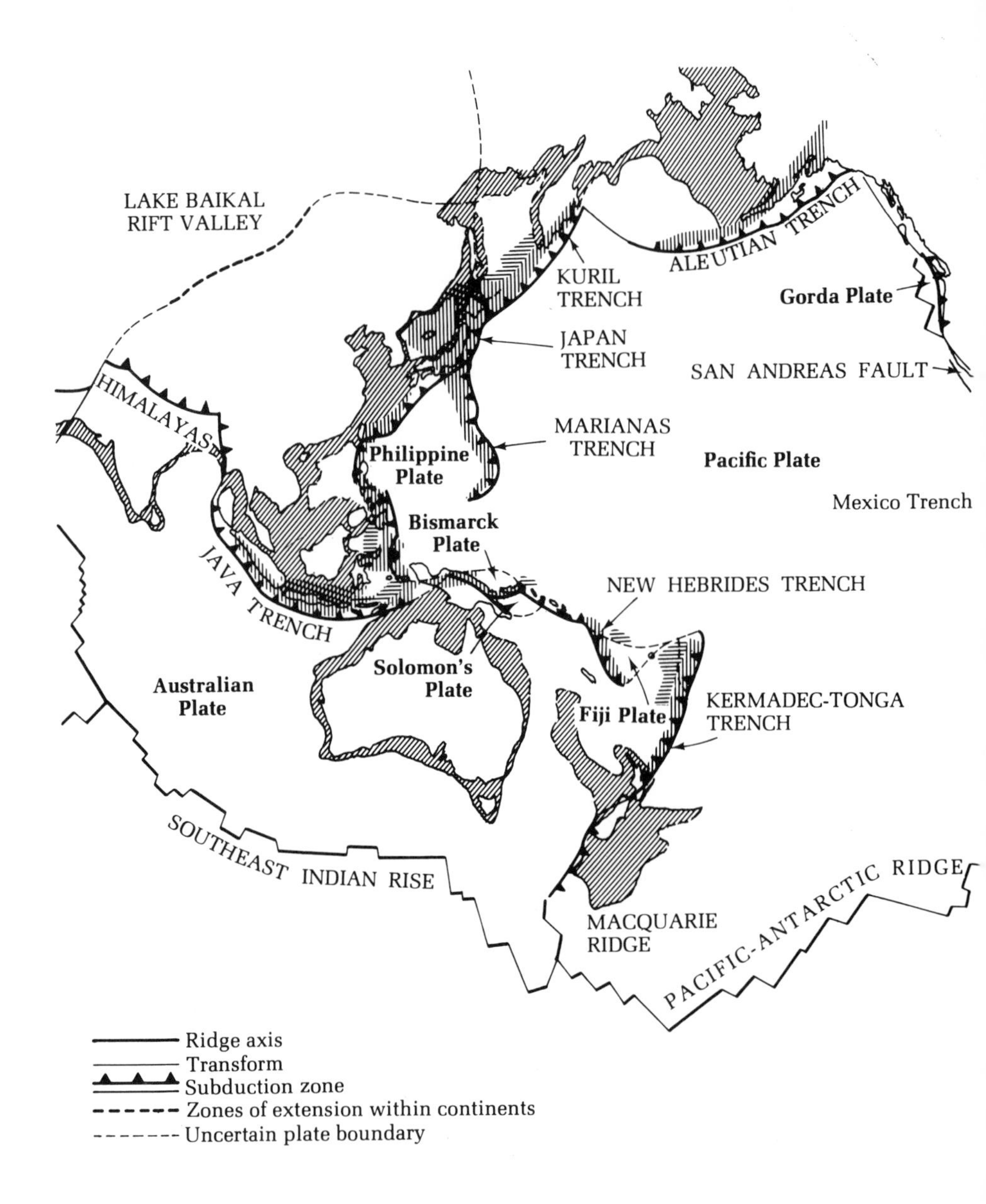

FIGURE 7.1 The Interlocking Network of Lithospheric Plates Which Form the Earth's Outer Shell. The rigid plates are in constant motion relative to one another. There are three types of boundaries or edges to the plates: (1) ridge axes from which the plates move apart and along which lava wells up to form new ocean floor crust; (2) transform faults along which plates slide past each other; and (3) subduction or sinking zones where plates collide and one plate edge is forced to dive beneath the leading edge of the other. Triangles show the leading edge of the plate. (From "Plate Tectonics," John F. Dewey. Copyright © May 1972 by *Scientific American*, Inc. All rights reserved.)

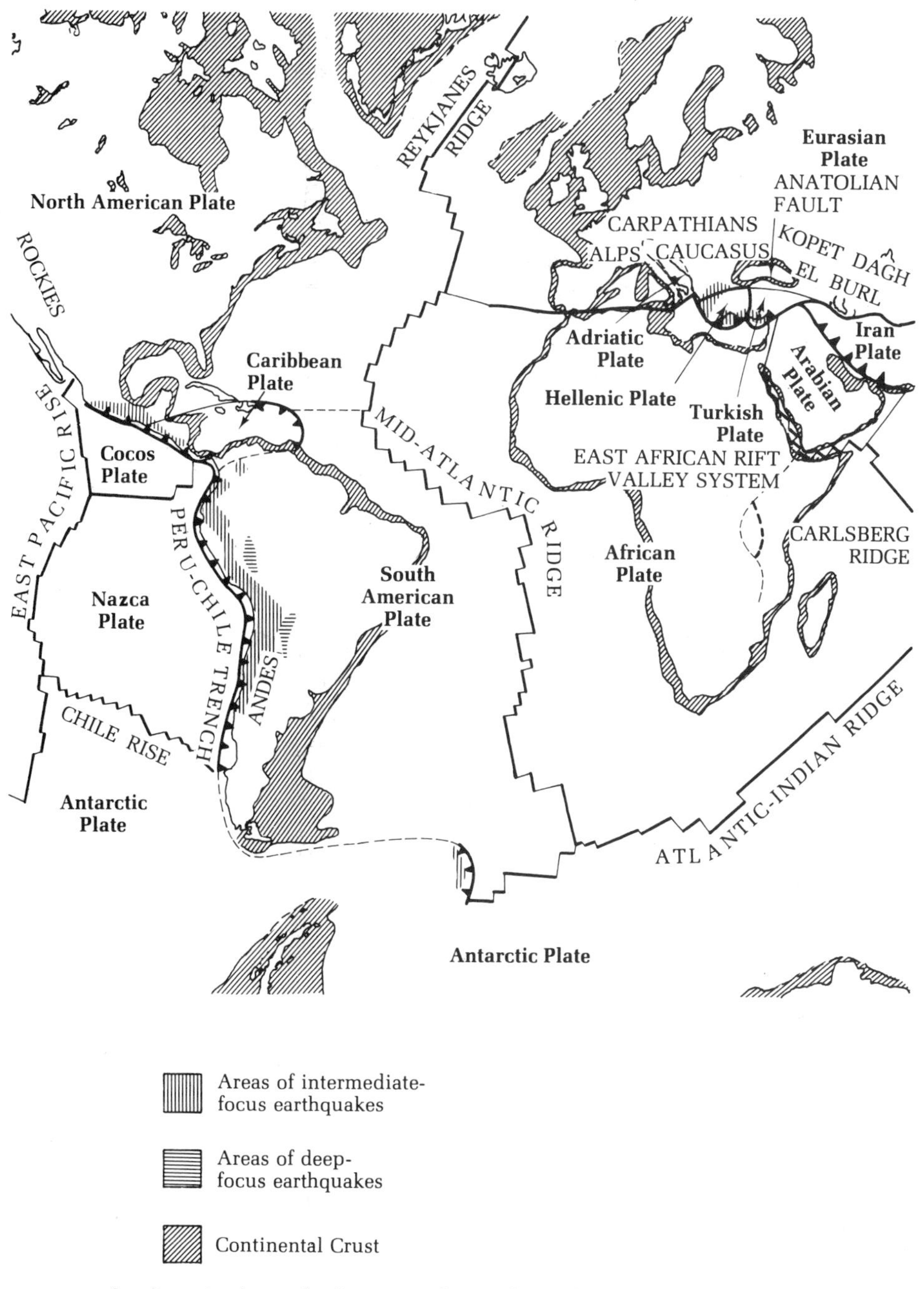

as the San Andreas fault zone, shows lateral or sidewise movement. The deep-ocean-trench subduction areas and the major compressed mountain range areas are both under compression. When earthquake zones are plotted on a map, they coincide with plate boundaries. It is now clear that as the tectonic plates move, motions of various kinds along the different plate boundaries result in the production of earthquakes. The most recent information about plate tectonics suggests the following pictures of the three kinds of plate boundaries.

Mid-oceanic-Ridge Spreading Centers and Magnetically Striped Rocks on the Ocean Floor Along the mid-oceanic ridge axes, the plates separate and move away from each other. As the plates depart from the spreading center, faulting and volcanic outpouring take place which build up mountains and escarpments along its length. Such spreading centers show high earthquake incidence, lines of volcanoes, magnetically striped rock bands symmetrically arranged on both sides of the spreading axis, heat flow several times greater than elsewhere on the crust, and topographic uplift. (See fig. 7.4.) Happily, such a plate boundary is exposed at the surface in Iceland where the Mid-Atlantic Ridge has risen above the water by rapid growth (See figs. 7.2a and 7.2b.) Iceland is cut by the rift zone which lies along the center of the ridge, with fault

FIGURE 7.2a Surtsey. A submarine volcano erupted in the ocean on November 14, 1963, about a dozen miles southwest of Iceland, forming the new island of Surtsey. The island is now more than one square mile and some of the craters are still steaming. This volcano formed astride the Mid-Atlantic-Ridge spreading center as the Eurasian and North American plates drifted apart. Spreading-center volcanoes such as Surtsey and Kirkjufell, which formed this year (fig. 7.2b) are relatively quiet compared to the highly explosive and dangerous type which form near the deep-sea trenches where lithospheric plate edges are underthrust to be resorbed in the mantle. (Photograph Courtesy of Icelandic Airlines.)

FIGURE 7.2b Kirkjufell. A volcanic eruption began January 23, 1973, on the island of Heimaey in the Westman Islands, about 10 miles southwest of Iceland. It forced the evacuation of Heimaey's 5200 residents and covered more than one-half the buildings in the photograph with ash and lava. The new volcano, named *Kirkjufell,* is astride the Reykjanes Ridge which is the northern extension of the Mid-Atlantic Ridge. (Photograph Courtesy of Icelandic Airlines.)

steps making up the sides of the rift valley. The many cracks of the rift zone have filled with rising lava, which cooled to form tabular, roughly vertical, injected rock bodies called **dikes.** Over 200 volcanoes are also found along the floor of the rift valley; many are still active. Heat flow is very great as evidenced by subsurface tapping of hot water to heat the city of Reykjavic. The great open tensional cracks along the Icelandic rift are numerous and clearly visible. (See fig. 7.3.) When up-and-down movements occur along these cracks, fault block ridges are produced; a series of them are arrayed like giant steps along the valley sides. Although lava eruptions fluctuate over short periods, there appears to be some constancy over great periods of time.

The separation of plates along the mid-oceanic-ridge boundaries produces tension and reduces the confining pressure in the mantle rock (asthenosphere) below the lithosphere (brittle outershell). Some melting of asthenosphere mantle material results, and the molten rock rises, pushing up the ridge area. The lava rises further to fill the tensional cracks of the rift zones and cools to form new basaltic crust. Thus, at the ridge axes, new surface area is created along the trailing edges of the plates by volcanic discharge as the plates move apart. The mid-oceanic-ridge spreading centers are not continuous over great distances

FIGURE 7.3 Thingvellir Graben. The Mid-Atlantic Ridge rises to the surface in Iceland, showing its great central rift zone with tensional structures; these are produced as the Eurasian and North American plates drift apart. Here the great rift valley referred to as the *Thingvellir Graben* can be seen in the southwest part of Iceland. The tensional tears along the rift provide conduits along which the lava erupts to form new crust. (Photograph Courtesy of Icelandic Airlines.)

but are divided into offset segments. The segments are connected by transform faults, which we shall turn to shortly.

Measurements of variations in the earth's magnetic field done with magnetometers have revealed that such variations are considerably greater over the ocean floor than over the less magnetic continental rocks. About 1960, in the midst of new speculation on drift, workers from the Scripps Institution of Oceanography announced that the ocean floor off the North American west coast showed an unusual pattern of variations in the magnetic intensities of the rocks underlying it. A series of parallel rock stripes formed the pattern; they ran roughly north-south and were broken and offset along roughly east-west lines at right angles to the stripes. This pattern was unexplained until 1963, when F. J. Vine and D. H. Matthews suggested that if lava boiled out along the center of the mid-oceanic ridges, the iron minerals in it would become magnetized in the direction of the earth's magnetic field present at that time. The newly generated, magnetically imprinted rock would then be carried away from the ridge

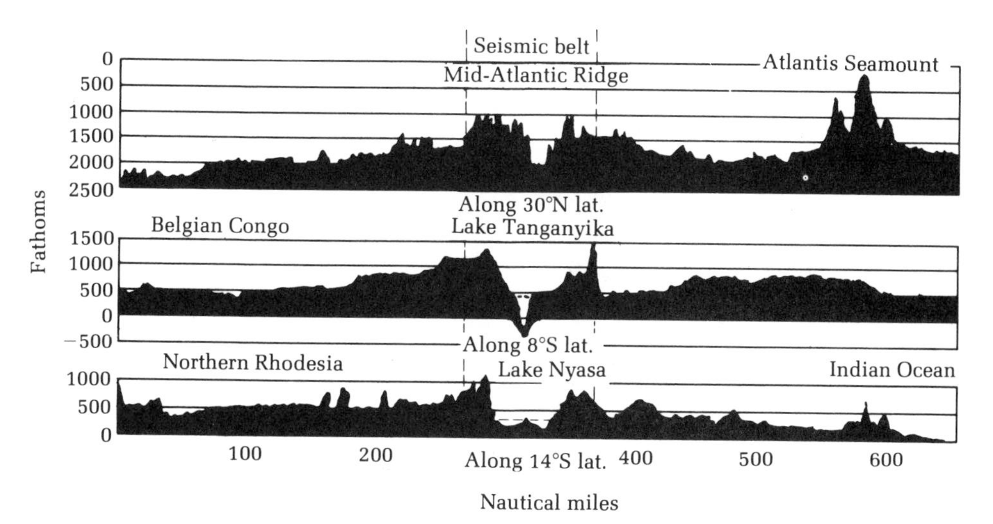

FIGURE 7.4 Profile across the Mid-Atlantic Ridge and Comparable Profiles across the East African Rift (Fault) Zone at Lake Tanganyika and South at Lake Nyasa. The Mid-Atlantic Ridge has long been known, since Iceland, St. Helena, the Azores, and other islands are surface expressions of it. The islands are mostly basaltic rock such as that formed in recent eruptions in Iceland. Ancient rocks found on St. Peter and St. Paul Rocks which lie on the ridge suggest that they may be original mantle material formed at the time of earth origin and later thrust to the surface by the forces that built the ridge. A mid-oceanic ridge cuts inland to join the East African rift valleys. They are not as symmetrical as the rifts down the center of the oceanic ridges but are similar in their high plateaus, suggesting uplift as in the mid-oceanic ridges. (Reprinted, by permission, from Bruce C. Heezen, *The Deep Sea Floor*. New York: Academic Press, 1962.)

centers at uniform rates of speed. This theory was also independently proposed in 1964 by Morley and Larochelle. It was an outgrowth of the original Holmes convection current hypothesis of 1928 which was expanded by Hess in 1960 and by Dietz in 1961. Vine and Matthews searched for evidence of polar reversals in the magnetized sea-floor rocks to correlate with those on land, which had already been demonstrated. They believed that as the sea floor spread, it would carry the rock away from the ridge center in both directions. As new material erupted at a later date along the ridge, it would in turn be magnetized, capturing the possibly reversed magnetic field. If this process had taken place repeatedly, they hoped to find alternating strips of rock displaying normal and reversed magnetism arranged symmetrically about the spreading center. (See fig. 7.5.)

At the same time as Vine and Matthews' suggestion, J. R. Heirtzler and a group from Columbia's Lamont-Doherty Geological Observatory and the U. S. Naval

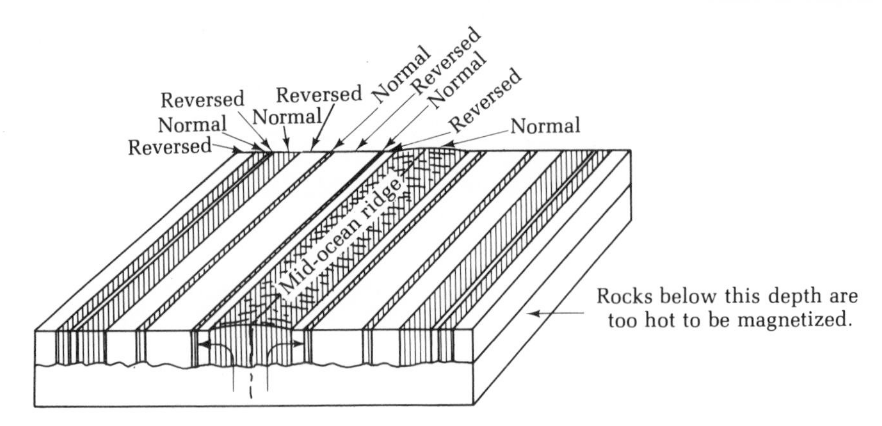

FIGURE 7.5 Magnetization of New Sea Floor along the Mid-oceanic Ridges. Lava erupts along the crest of the mid-oceanic ridges and becomes magnetized, imprinting the earth's magnetic polarity for that interval. The spreading sea floor carries the rock away from the ridge center in both directions. New rocks erupted along the ridge at a later time would in turn be magnetized, possibly capturing a reversed magnetic field. This continuing process leads to alternately magnetized bands of rock showing normal and reversed polarity arranged symmetrically about the spreading center.

Oceanographic office were towing sensitive magnetometers behind an ocean research vessel over the Reykjanes Ridge, which is part of the Mid-Atlantic Ocean Ridge lying just south of Iceland. They found that magnetic variations were indeed laid out in stripes. Additionally, the magnetic stripes were distributed parallel to the axis or center of the ridge in a symmetrical way such that the sequence of stripes on one side of the ridge formed a mirror image of the pattern on the opposite side of the ridge (see figs. 7.6a and 7.6b).

Vine and Matthews discovered an important key to sea-floor spreading. They showed that the ratios of the width of the magnetically striped rocks of opposed polarity on the sea floor are identical to the ratios of polarity intervals observed earlier in young rocks on land. This ratio is the key to interpreting spreading rates. The work of Allan Cox and R. R. Doell during the 1950s in establishing a time scale of polar reversals from younger magnetized rocks of the continents (see fig. 6.6) really provided the essential base on which Vine and Matthews worked. The exact dating and counting of magnetic reversals recorded in rocks on land does not extend back more than about 6 million years. Magnetic reversals in older rocks have been found, but they cannot yet be dated or counted; all older reversals are therefore based on evidence from the sea floor.

Shortly after this, F. J. Vine and J. Tuzo Wilson showed that the symmetrical magnetic pattern from the Reykjanes Ridge south of Iceland closely matched the pattern of reversals found on the sea floor off the North American coast (see fig. 7.7). The ratio of the time scale of polarity reversals derived from lava flow rocks on land is the same as the ratio of the successive magnetic anomaly strips parallel to the mid-oceanic ridges.

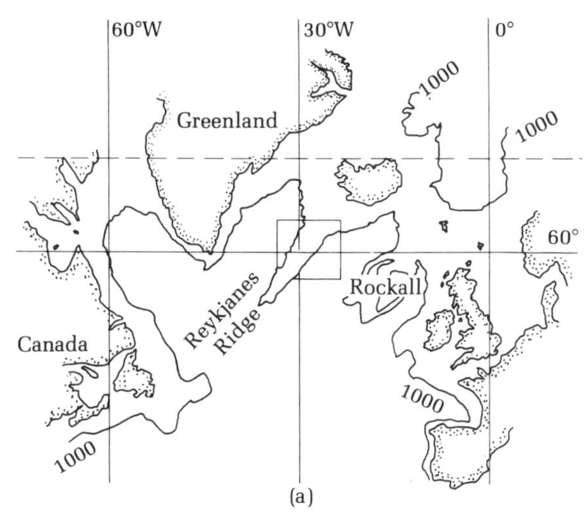

FIGURE 7.6a The Location of the Reykjanes Ridge Just South of Iceland. This area is shown enlarged in fig. 7.6b. The 1000- and 500-fathom submarine contours are shown for the Rockall Bank. (Reprinted, by permission, from J. R. Heirtzler, X. Le Pichon, and J. G. Baron, "Magnetic Anomalies over the Reykjanes Ridge." *Deep-Sea Research* 13. Courtesy of Microforms International Marketing Corp., Elmsford, N. Y.)

Vine and Wilson showed that the magnetic anomaly strips, which parallel the crests of the mid-oceanic ridges, occur in successive widths which are in the same ratio as the ratio of times in the polarity reversal time scale based on terrestrial lava flows. These two apparently identical ratios were reinforced by Opdyke's group in 1966; Opdyke provided a third identical ratio derived from deep-sea core samples which showed weak, but measurable, magnetization. All the cores displayed the same succession of normal and reversed magnetization at the same depth, in perfect agreement with the first two ratios. Thus, three separate and independent sets of complex data involving great geologic time, wide geographic distance, and successive sediment layers were found correlative. Tuzo Wilson has remarked that these data comprise a revolution in earth science in that "three different features of the earth all change in exactly the same ratios. These ratios are the same in all parts of the world. The results from one set are thus being used to make precise numerical predictions about all the sets in all parts of the world."

The successful matching between ridges prompted the reexamination of all available magnetic-ocean-floor records, and by 1968 J. R. Heirtzler and his associates revealed that the same pattern of magnetic reversals could be identified on almost all of the mid-oceanic ridges, although the width of the pattern varies from ridge to ridge because of different spreading rates. Since reversals of the last few million years were accurately dated, they were able to use those data in extrapolating backward. They therefore constructed a time scale for magnetization of the sea floor for the Cenozoic (the past 65 million years); this scale facilitated the interpretation of spreading rates. Other kinds of

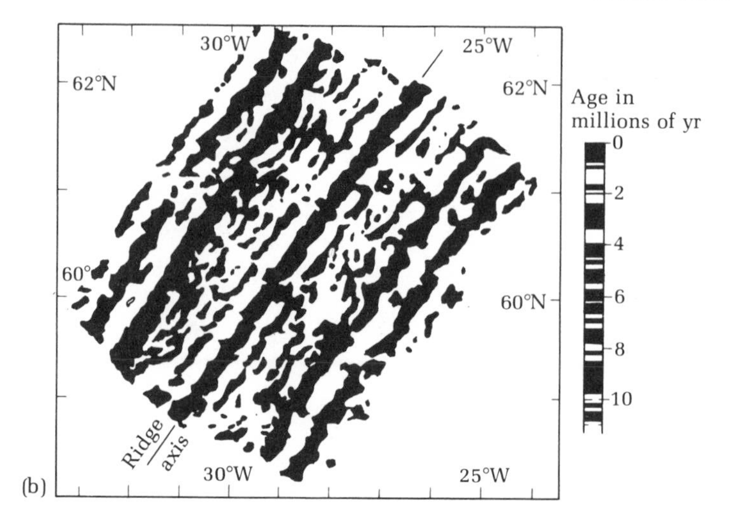

FIGURE 7.6b The Magnetic Variation or Anomaly Pattern of the Reykjanes Ridge, Which Is Part of the Mid-Atlantic Ocean Floor Ridge Just Southwest of Iceland. It is based on an interpretation of a magnetic survey by Vine and Matthews. The black stripes are inferred to represent rock on the ocean floor which shows positive magnetization (the same as the earth's present magnetic field). The white stripes between the black are inferred to correspond to rock beds with negative magnetization (produced when the magnetic poles were reversed). The youngest rocks lie at the ridge center, and increasingly older rocks are found in both directions away from the center. (Reprinted, by permission, from J. R. Heirtzler, X. Le Pichon, and J. G. Baron, "Magnetic Anomalies over the Reykjanes Ridge." *Deep-Sea Research* 13. Courtesy of Microforms International Marketing Corp., Elmsford, N. Y.)

independent evidence were not then available to test Heirtzler's prediction of spreading rates so far back in geologic time. Shortly afterward, the J. O. I. D. E. S. (Joint Oceanographic Institutes Deep Earth Sampling Program) provided sediment cores from the South Atlantic, dated by microfossils, that increased in age with increasing distance away from the ridge crest. The cores implied a spreading rate of 2 cm/year (4/5 inch); this figure agreed closely with that predicted by Heirtzler from his paleomagnetic data. (See fig. 7.8.)

Because the spreading centers produce elongate fissures, much of the lava extruded accumulates to form long volcanoes rather than the familiar cone-shaped ones. It is thought that if spreading occurs slowly, the rising lava has more time to cool and becomes viscous; thus it solidifies quickly when erupted and does not flow away from the axis, giving rise to high, steeply sloping ridges. When spreading occurs rapidly, the lava is brought to the surface quickly in a hot, more freely flowing state which allows it to move away from the spreading center and produce lower ridges with gently sloping flanks.

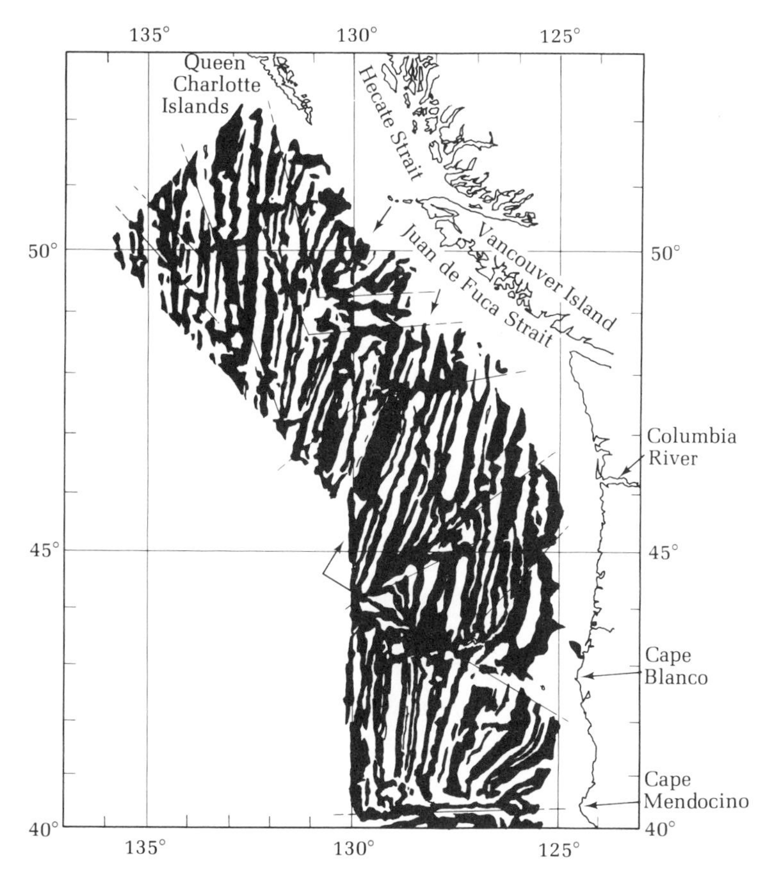

FIGURE 7.7 Magnetic Pattern in the Ocean off the North American Coast. The black is inferred to represent positive magnetization, and the white areas between are thought to be rocks of reversed magnetization. The **straight lines** breaking and offsetting the pattern are faults. The **arrows** point to short segments of oceanic ridge with the pattern of normal and reversed magnetic stripes parallel to and symmetrical about the ridges. [Reprinted, by permission, from F. J. Vine, "Spreading of the Ocean Floor: New Evidence." *Science* 154 (16 December 1966): 1405–15. Copyright 1966 by the American Association for the Advancement of Science.]

Spreading centers that open quickly at rates from 2 to 5 inches per year build ridges less than 1500 feet high; in contrast, when spreading is less than 2 inches per year, high ridges result. The new crust at the spreading center is elevated above the surrounding sea floor as a mid-oceanic ridge by the expansion of the mantle below, probably due to heating. As the new crust moves away from the spreading center, the mantle below it cools, resulting in contraction and sinking. This is supported by measurements of heat flow which show

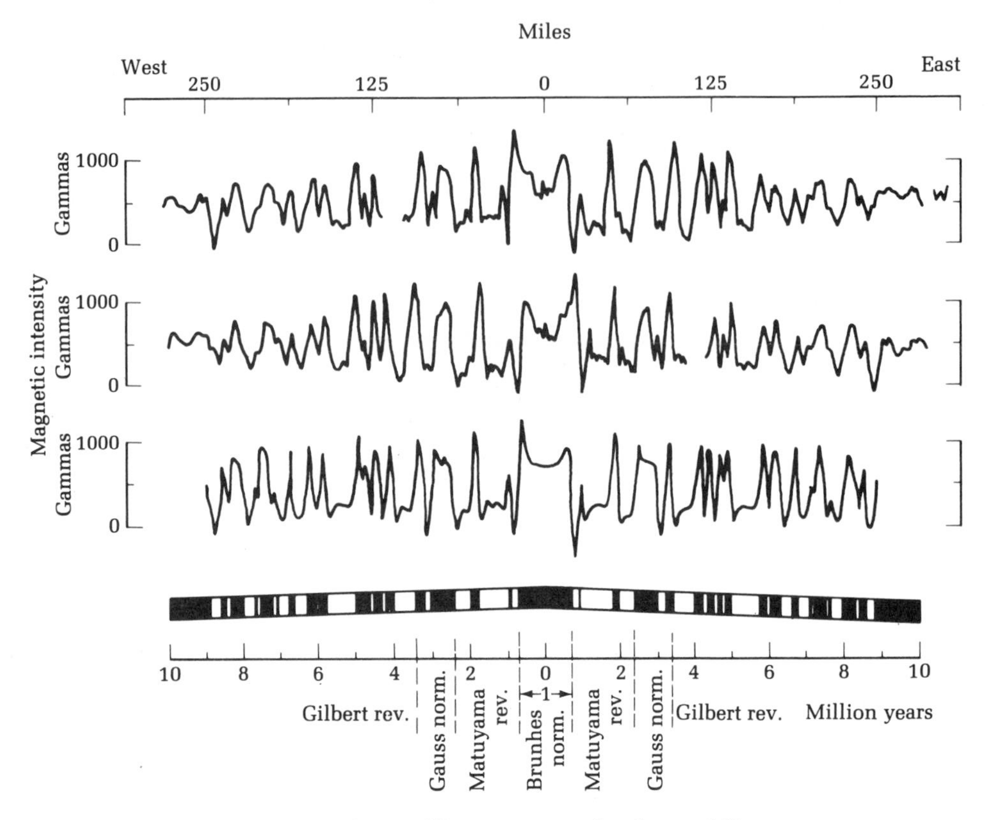

FIGURE 7.8 Magnetic Profiles Compared. The middle curve is a magnetic profile across the Pacific-Antarctic ridge (the southwestern part of the East Pacific Rise at 50°S, 120°W). The strength of the magnetic field is laid out showing distance from the mid-oceanic ridge center. Distance scale is at the top. The profiles run east-west. The top curve is the same as the middle but has been reversed (east is on the left). Comparison of these two curves reveals an almost mirror image symmetry about the ridge axis. The lower curve was computed from a model that would be produced by alternating strips of crust bearing normal and reversed magnetization. The spreading rate is almost 2 inches per year. Normal crustal blocks are shown in black. The ridge center is composed of a wide, normally magnetized, crustal block. (Reprinted, by permission, from W. C. Pitman III and J. R. Heirtzler, *Science* 154 (2 December 1966): 1164–71. Copyright 1966 by the American Association for the Advancement of Science.)

decreasing values with increasing distance away from the mid-oceanic ridges. It is estimated that the crust sinks rapidly for the first few million years after it forms and spreads, perhaps 300 feet per million years, but gradually declines to about 50 feet per million after about 50 million years. The plates depart from the mid-oceanic-ridge spreading centers at rates which vary from approximately 3 to 15 miles per million years in different areas, but the rate appears to be fairly constant in any given area. The Mid-Atlantic Ridge is a steeply sloped, narrowly elevated structure produced by low spreading rates. Each plate along the North Atlantic part of the ridge moves at only ½ inch per year. In contrast, the East Pacific Rise is wide and gently sloping, which is the result of fast spreading along its central rift zone; here each side of the spreading center departs by more than 3 inches per year. Sinking is fastest near the rift center and slows with distance; thus the mid-oceanic ridges have a concave profile skyward. (See figs. 7.9a and 7.9b.) The spreading rates for the major ridges are summarized in table 7.1.

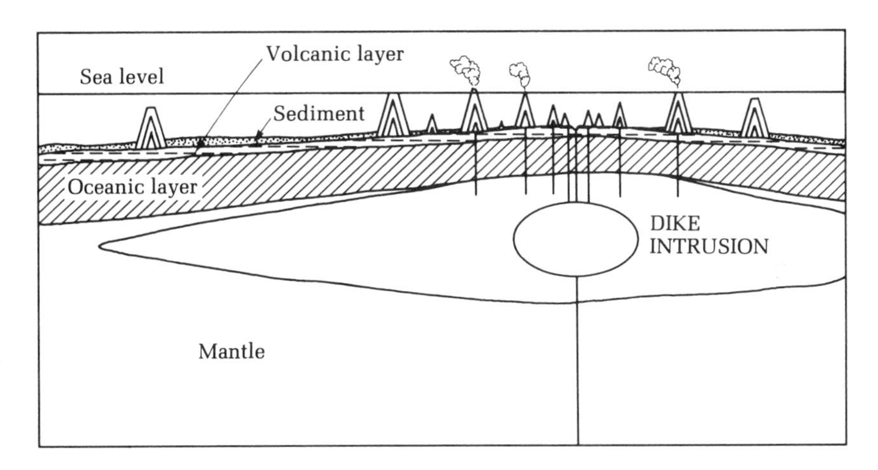

FIGURE 7.9a Rapid Spreading at a Mid-oceanic Ridge. Here the plates depart quickly, forming a thin layer of volcanic rock. Because the plates move rapidly, the volcanoes which usually begin to grow near the ridge axes are carried far from the spreading center before they reach the zone of wave action. There they are planed flat on top to form a seamount or guyot which becomes capped by sediment. Only if a volcano rises fast enough (2½ miles high in 10 million yr) will it emerge as an island. In the South Atlantic, St. Helena Island formed in this way. (Reprinted, by permission, from H. W. Menard, *Journal of Geophysical Research* 74, p. 4833, 1969, copyright by American Geophysical Union.)

Table 7.1 Comparison of Spreading Rates around the World.

Spreading Center	Spreading Rate per Year in Inches (Both Plates)
North Atlantic Ridge	1¼
North Indian Ocean	1½
South Atlantic Ridge	2
South Indian Ocean	
North Pacific Ocean	3¾
Pacific-Antarctic Rise	5
East Pacific Rise	7

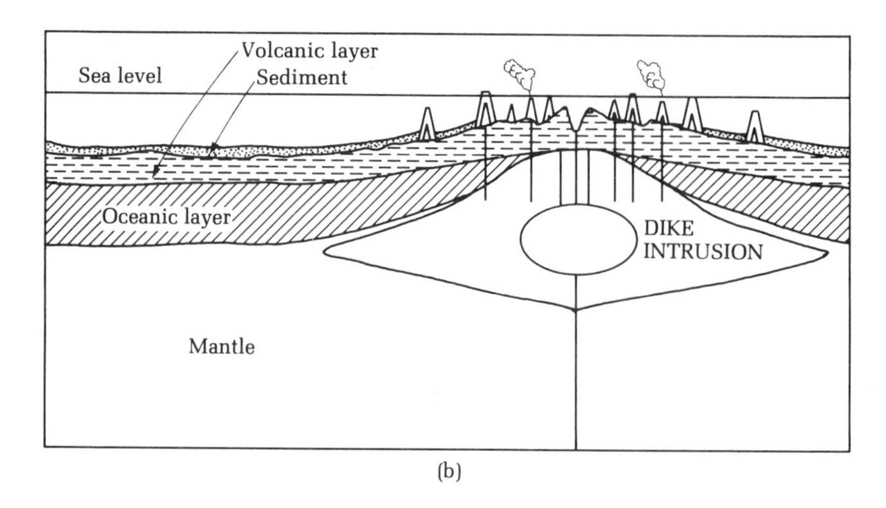

(b)

FIGURE 7.9b Slow Spreading at a Mid-oceanic Ridge. In contrast, slow spreading gives rise to a thicker volcanic rock layer. The growing volcanoes, in the case of a slowly moving plate, reach the surface when they are still close to the plate edge. A flat platform is cut across their tops by wave action; they then submerge as the plate continues to depart the spreading center, subsiding at the rate of 300 to 50 ft per million years, rapidly at first, but slowing to the lower figure after about 50 million years. (Reprinted, by permission, from H. W. Menard, *Journal of Geophysical Research* 74, p. 4833, 1969, copyright by American Geophysical Union.)

Most continents are riding on plates in which the spreading-center, plate boundaries are in the ocean and do not cut the continents, but such separation of a continental landmass is presently resulting in the opening of the Red Sea and the Gulf of California. (See fig. 7.10.) It is quite amazing that in spite of the great length of time that the continents have been separated, they still in many cases can be reconstructed along the coasts first created when a spreading center operated under the landmass of which they were a part. The use of the computer in the matching of coastlines by Bullard and others has produced remarkably close fits among the pieces of the jigsaw puzzle, as we saw in chapter 3.

The spreading centers are not fixed in geographic position. In spite of the fact that new crust is generated there such as to produce symmetrical or mirror-image magnetic patterns, ridge flanks, and mountain ranges, which all suggest a fixed or stationary position, there is evidence that spreading centers often move. Manik Talwani of the Lamont-Doherty Geological Observatory believes that all spreading centers move. In order to produce symmetrically patterned bands of rock on both sides of the spreading center, it is required that the spreading center migrate at exactly one-half the speed at which the plates separate. If this requirement were not met, then no mirror image symmetry would be produced across the spreading center, and the magnetically striped rock bands would become jumbled (see fig. 7.11).

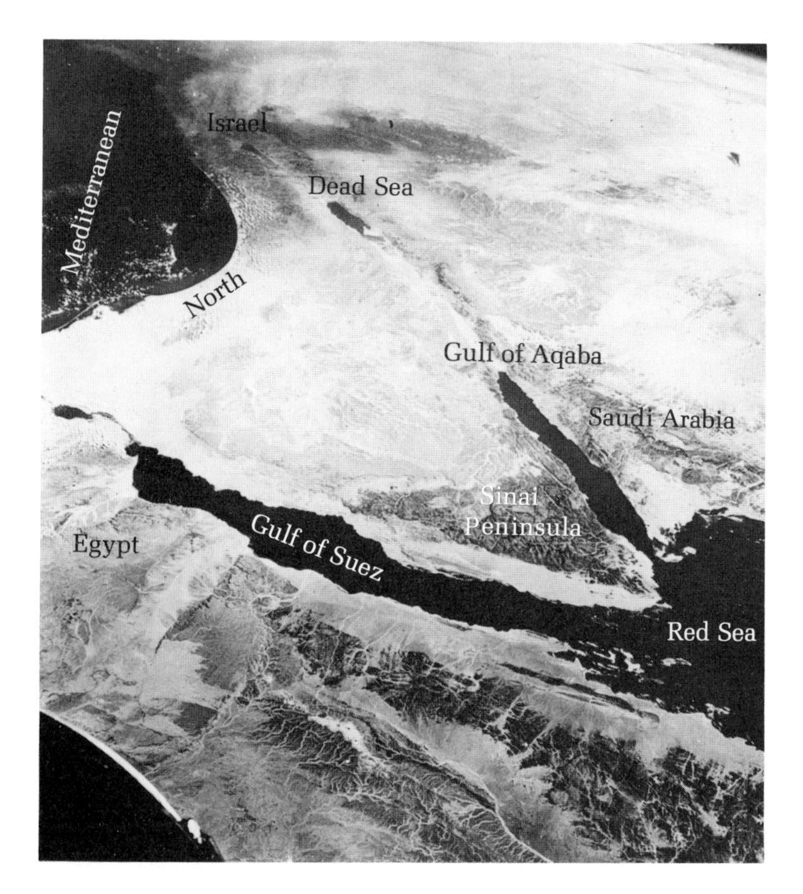

FIGURE 7.10 The Red Sea and Its Fingers, the Gulfs of Suez and Aqaba. The southeastern edge of the Mediterranean Sea is at upper left. During the past 5 million years, Africa has separated from Arabia, forming the Red Sea. The Gulf of Aqaba occupies a depressed fracture zone along which the Dead Sea is located to the north. Similarly, the Baja Peninsula appears to have split from the American mainland to form the Gulf of California. Both of these areas appear destined to become oceans in the future. They may provide us with a model of the Atlantic Ocean during its early stage of development about 175 million years ago. (Gemini XI Photograph Courtesy of NASA.)

Apparently the Chile Rise off the South American coast and the East Pacific Rise are adjacent spreading centers without a subduction or crustal sinking zone between them. In a subduction or sinking zone the crust plunges downward to be destroyed by resorption in the mantle. Since new plate material is being added continually on the inside edge of each rise or spreading center, we must expect that the crust between them would be buckled into a mountain range or pushed downward into a trench. Since no trench or mountain range is present, it must be assumed that one or both spreading centers are moving such as to accommodate the newly generated crust. The same situation must exist

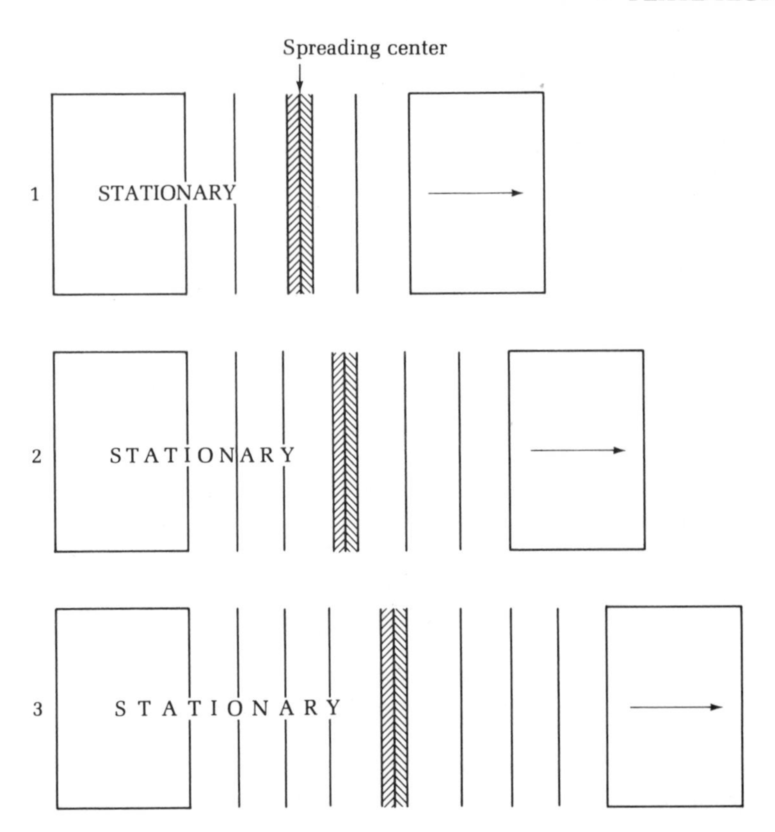

FIGURE 7.11 A Moving Spreading Center. The center produces a symmetrical pattern of magnetized rock bands if it migrates at one-half the rate at which the plates move apart. The upwelling lava is extruded and becomes magnetized, recording the magnetic field of the earth prevailing at that time. Changes in the earth's magnetic field are captured during successive eruptions as spreading continues. In the above diagram, the left-hand plate is stationary while the right-hand plate moves to the right. In fig. (1), molten rock is injected along the spreading center, cools to hardness, and is split down the center. In fig. (2), the succeeding injection of hot material is emplaced between the two halves of the earlier rock mass. The new rock body has therefore been emplaced one-half the width of the earlier mass farther from the fixed plate than was the earlier mass. In fig. (3), the new hardened mass has split, and another rock mass has been emplaced a whole width farther from the stationary left-hand plate. (After Menard, 1969.)

between the Mid-Atlantic Ridge and the Carlsberg Ridge in the Indian Ocean; the two do not have a subduction zone or a folded mountain range between them, which again suggests that one of them must be moving. It is now believed that some spreading centers have migrated great distances so that one rise may become another elsewhere.

If the ocean floor crust is generated at a spreading center and moves outward, then we should expect the sediment which accumulates on the ocean floor to vary in thickness in relation to distance from the spreading centers. Rates of sedimentation have been compiled for a great number of localities on the ocean floor and found to vary from ½ to 2 centimeters (1/5 to 4/5 inch) per thousand years. If sediment has been accumulating at the rate of the smaller figure, we should expect to find about 5 miles of sediment. Instead, studies of penetrating shock waves reveal an average thickness of less than one mile on the deep-ocean floor. This fact was known before the concept of sea-floor spreading was put forward. It has led some geologists to believe that the oceans are young features of the earth. Recent measurements of the thickness of sea-floor sediments indicate that the sediment gets both thicker and older at progressively greater distances from the mid-oceanic-ridge spreading centers. This is certainly a strong argument in favor of a spreading sea floor.

Subduction or Sinking Zones The second type of plate boundary is the subduction zone along which the leading edge of one plate is thrust under the leading edge of another. Such zones are marked by deep-sea trenches and young mountains. It is now believed by X. Le Pichon that if the plates move together and shorten the crust by more than 2 inches per year, the lithosphere is deformed to form trenches; if the plates move at a slower rate, major mountain ranges are formed by folding and overthrust faulting. Recall that the tectonic plate is about 40 to 90 miles thick, made up of crust and uppermost mantle. This cold, brittle, lithospheric plate meets increasing resistance as it is thrust down into the soft zone of the mantle called the asthenosphere (fig. 7.12).

Where the plate descends, it is curved and under tension, bending elastically. Cracks also form, producing great fault blocks which lie parallel to the trench. The growing plate is clean as it departs the mid-oceanic-ridge spreading center, but in its long, slow journey across the sea floor to its destructive destination by downturning in a trench, it acquires both volcanic and ocean sediments on its surface. Such surface accumulations of a plate appear to be represented by the rocks of the California coast ranges, which are typical ocean-floor sedimentary and igneous rocks. They were acquired during an earlier time when a subduction zone lay along California's western margin. The eastward-moving, Pacific Ocean floor plate plunged under in that subduction zone, and these encrusting, more buoyant, telltale, ocean-floor rocks were pushed up to form the coast ranges. The plate meets little resistance at shallow depths in the asthenosphere, but as it dives deeper, it appears to meet greater resistance, and it heats up from both friction and conduction of heat from the hot mantle. Some evidence suggests that pieces of the descending plate break off and plunge into the mantle at faster rates. Since the various rock-forming minerals of the lithosphere melt at different temperatures, those that melt at low temperatures separate at about 60 to 100 miles and rise to yield the silica-rich lavas that form the volcanoes of the island arcs bordering the deep-ocean trenches. The volcanoes are strung out along the leading edges of diving plates which certainly suggests a connection between the two. The volcanic rocks of the subduction zones are less dense than those of the oceanic crust which are generated at the mid-oceanic ridges. Lavas are thought to be formed by the partial melting of oceanic crust and other lighter rocks carried down into the hot asthenosphere at the subduction zones. There are indications that the diving, oceanic, lithospheric plate is easily digested by subduction because it is

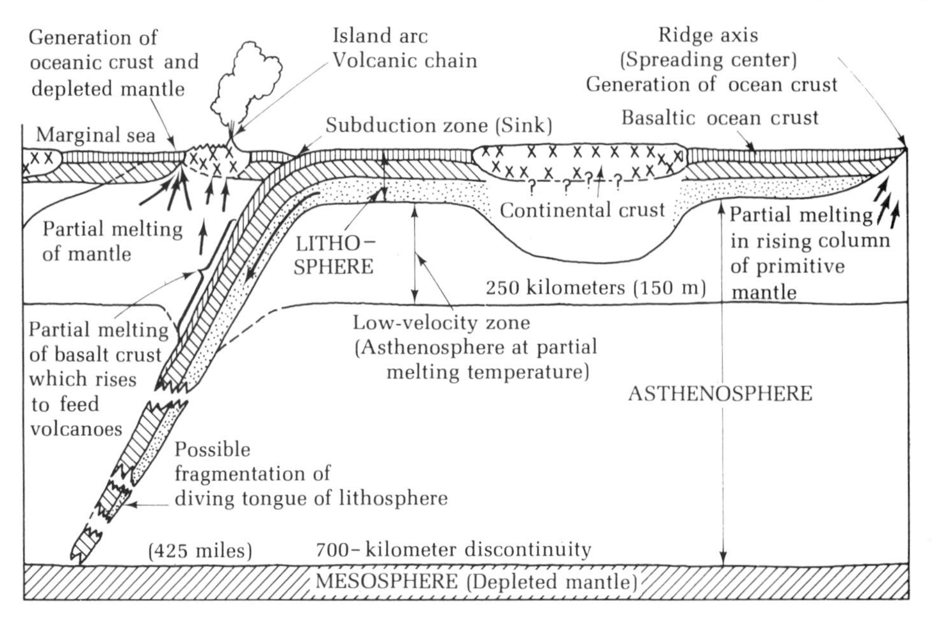

FIGURE 7.12 Lithospheric Plate in Schematic Cross Section. The solid lithosphere "floats" on the semimolten upper asthenosphere. The lithosphere, which is much thicker under the continents, is moving toward a subduction zone where the plate is forced beneath the leading edge of the opposing plate and plunges into the mantle. It is possible that pieces of the underthrust plate break off and plunge at faster rates. Rising basaltic lava is continually adding new ocean crust at the spreading centers. (From "Plate Tectonics," John F. Dewey. Copyright © May 1972 by *Scientific American,* Inc. All rights reserved.)

dense and thin, but a different phenomenon occurs when two thick, light, buoyant continental lithospheric plates move into a subduction zone and collide. Such a collision produces compression over wide areas and results in the formation of great mountain ranges such as the Alpine-Himalayan ranges. Certain rock types within these ranges called **ophiolites** are thought to be parts of the oceanic crust and mantle. We shall mention them in a later section. They are believed to lie along the line of collision between two continents; of course, such continental collisions destroyed the ocean which once lay between them. The Mediterranean Sea is thought to be a small remnant of the once great Tethys Sea which lay between Europe and Africa. When two continents collide, subduction or lithospheric depression is ended along the line of impact, and a new subduction zone probably forms elsewhere (fig. 7.13).

In some cases two plates may be brought in contact such that a continent lies at the edge of one and an oceanic trench bounds the other. The heavier oceanic crust is then pushed under the opposing continental margin. A modern example is provided by the eastern Pacific plate margin which is being thrust under the west coast of South America. This also occurs where the western Pacific plate edge dives under eastern Asia. Understanding of this phenomenon now

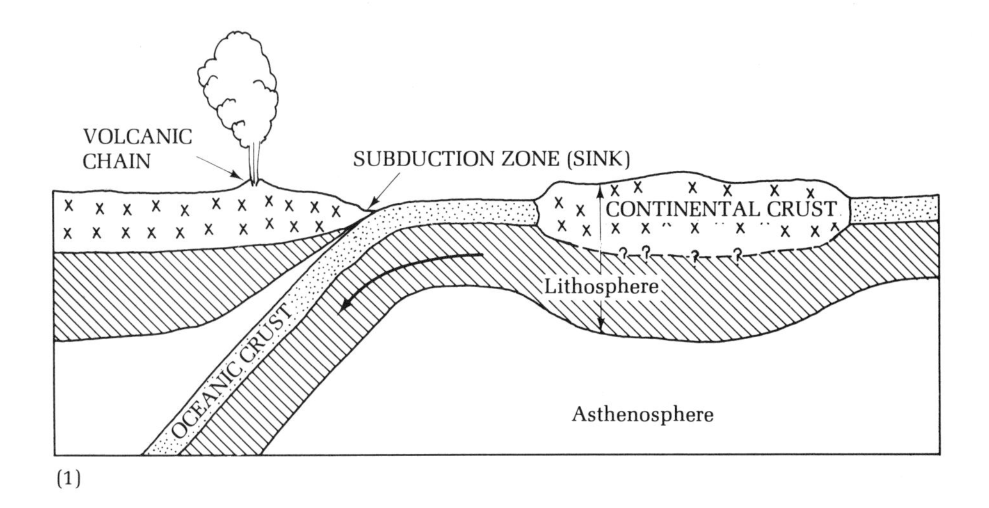

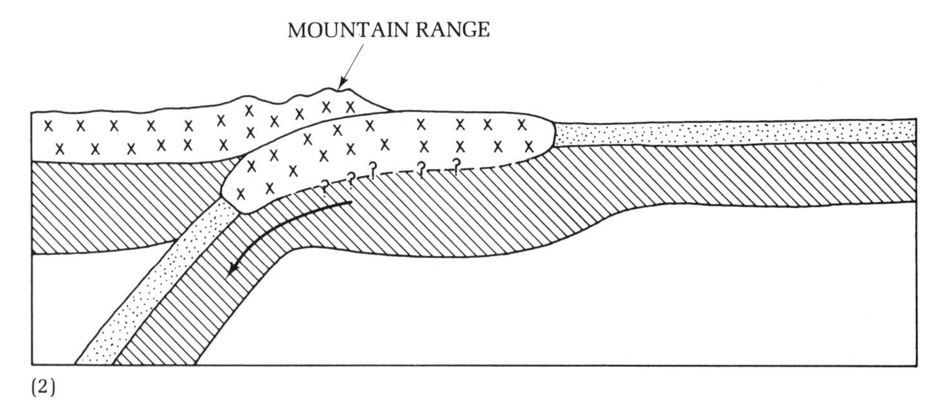

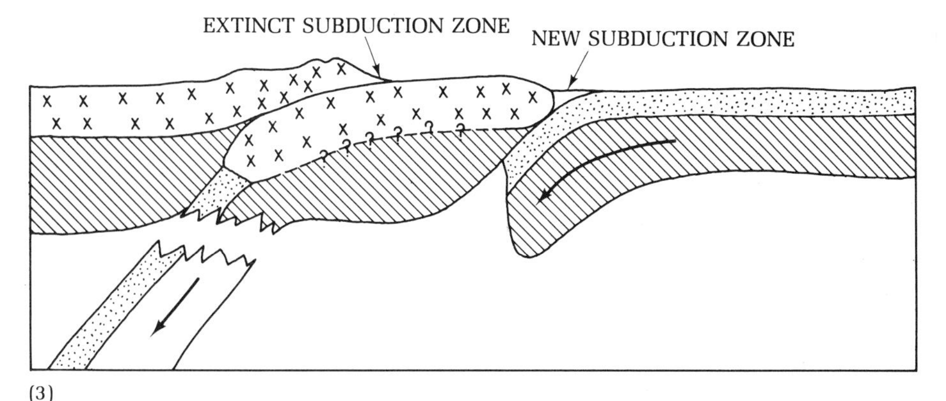

FIGURE 7.13 Collision of Two Continents. Two continents collide at a subduction zone (1). Because continental crust is lighter or more buoyant than oceanic crust, it is not dragged down into the asthenosphere but instead piles up to form mountain ranges such as the Alps and Himalayas (2). Following such a collision, the plunging lithospheric plate may be torn off and may sink. A new subduction zone may form later at a new site (3). (From "Plate Tectonics," John F. Dewey.)

allows interpretation of many ancient geologic structures such as the western or Cordilleran mountain ranges of North America which are believed to be a product of such collisions.

Volcanic activity on the earth may now be interpreted in terms of this new earth model. Two main types of volcanoes on earth are known to be distributed on a basis which fits the plate tectonics picture nicely. The first volcano type erupts low-silica lava which flows easily, resulting in no great explosive or violent activity. This type, because of its predictability, may be viewed from fairly close in such places as Hawaii and Iceland where they form great basalt flows. The second type of volcano erupts high-silica, viscous or poorly flowing lava which cools to form the lighter-colored andesitic kinds of volcanic rocks. Because this lava flows poorly, it erupts violently. The basaltic, free-flowing volcanoes are largely confined to the oceans where their lava is derived from about 40 miles deep in the mantle. Such eruptions are taking place today along the mid-oceanic ridges where the lava piles high enough in some places to reach the ocean surface and form mid-oceanic islands. An example is the recently erupted one just south of Iceland along the axis of the Mid-Atlantic Ridge. (See figs. 7.2a and 7.2b.)

In contrast to the basaltic volcanoes, the andesitic type forms from the lava and gas which arise as part of the ocean floor crust is dragged down and melted in the hot part of the mantle below the subduction zone. These volcanoes,

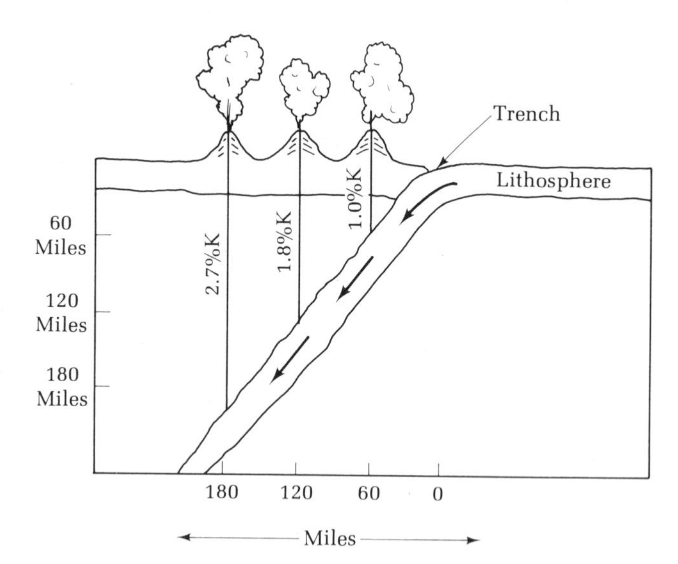

FIGURE 7.14 The Variation in the Proportion of Potassium to Silicon in Lavas Erupted in Island-arc Areas. The percentage of potassium (K) varies with distance from the trench. Volcanoes farthest from the trench are richest in potassium. Therefore, composition becomes a guide to depth at which melting occurred and may aid in determining the dip angle of the sinking tongue of lithosphere. It may also allow the location of ancient, extinct, subduction zones. (After Dickinson and Hatherton.)

which occur on the continents, are generally found along the edges of the tectonic plates near the trenches. As it sinks and passes into regions of increasing temperature and pressure, different products are driven off as liquids and gases at different levels. The earthquake depth in such volcanic regions is a clue to the depth of the top surface of a descending plate. Such estimates may allow the prediction of volcano composition, and, conversely, the presence of old volcanic andesite rock may reveal the presence of old subduction zones. It has been shown experimentally that the composition of the magma (molten rock) formed by partially melting rock depends upon the pressure or depth at which it melts. Melts which form at shallow depths (under low pressures) have little potassium in proportion to silicon. Melts produced at great depth (higher pressure) contain more potassium in proportion to silicon. The composition of lavas in active island-arc areas at the edges of continents display just such a

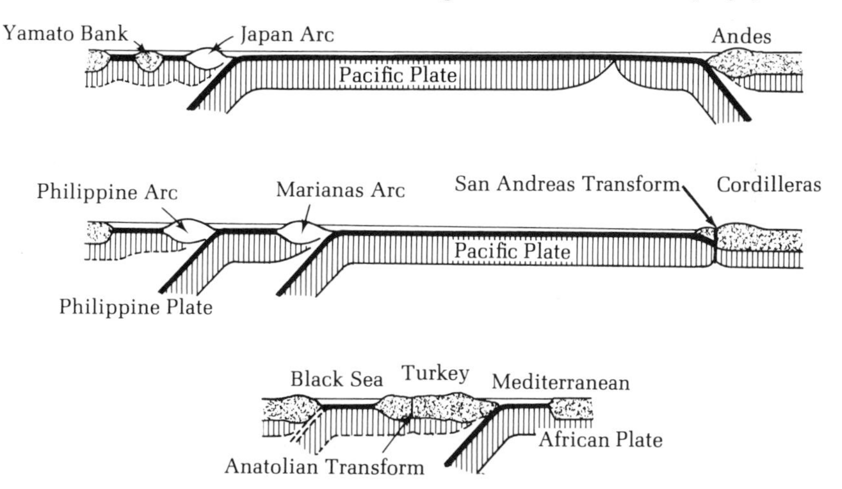

FIGURE 7.15 Examples of Modern Collisions between Plates in Which a Continent Lies at the Edge of One and an Ocean Trench Bounds the Other. (From John F. Dewey and John M. Bird, *Journal of Geophysical Research* 75, p. 2627, 1970, Copyright by American Geophysical Union.)

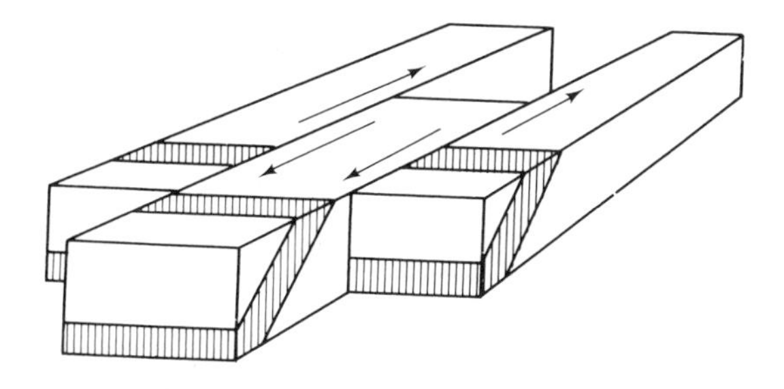

FIGURE 7.16 Lateral Faults. The horizontal movement has been parallel to the direction of the faults. Both faults shown are lateral; the fault on the right is left-handed, and the one on the left is right-handed.

pattern. William Dickinson and Trevor Hatherton (1967, 69) have shown that in island arcs and volcanically active continental edges there ". . . appears to be a relationship between the potash content of an andesite volcano and the depth of the seismic (Benioff) zone below that volcano." The volcanoes closest to the trench (farthest from the continent) are low in potassium; as volcanoes are sampled closer and closer toward the continent, the potassium content of the

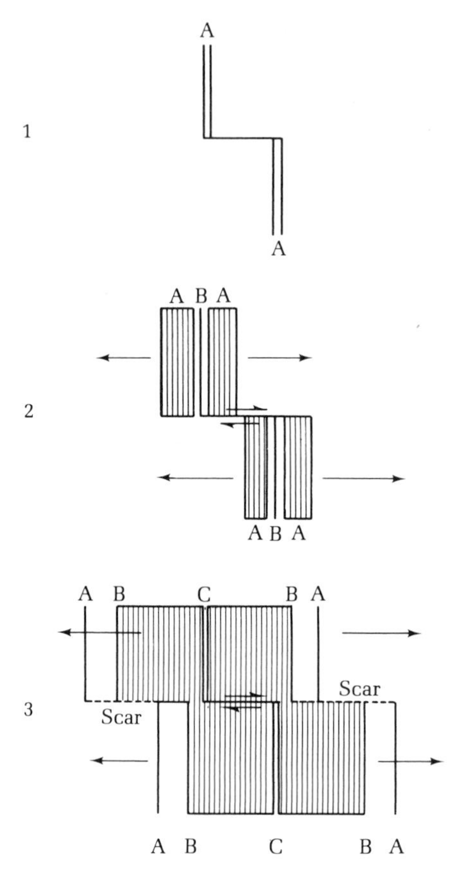

FIGURE 7.17a Map View of the Development of a Transform Fault Connecting Two Offset Segments of the Same Mid-oceanic Ridge. The double lines are a ridge along which spreading takes place and new crust is formed by upwelling lava. The single line connecting the double line is a transform fault. The three stages diagrammed are (1) beginning of the rift; (2) movement on the fault as indicated by arrows and the addition of new crust between lines B and B with continuing movement as shown by arrows. Displacement is now active only in that part of the fault which lies between two offset ridge segments. The extension of the fault beyond the active part is merely a scar representing a formerly active part of the fault; movement no longer takes place in this extinct area. The sense of movement on a transform fault is opposite that produced by simple lateral fault movement (see fig. 7.17b).

rocks rises. Evidently the descending crustal plate, inclined at a steep angle toward the continent, produces melts at different depths which rise to produce groups of volcanoes which are graded geographically in their potassium-silicon ratios (see fig. 7.14).

The rate of plate movement and sinking also becomes a guide to the expected volcanic activity in any area; an understanding of the direction and speed of plate movement may facilitate predictions about volcanic frequency and even possible reactivation of long-dormant areas (fig. 7.15).

Transform Faults The third type of tectonic plate boundary is the transform fault, along which plates move laterally relative to each other, but crust is neither created, as in the case of the mid-oceanic-ridge spreading centers, nor destroyed, as it is along the trench subduction zone boundaries. In 1965 J. Tuzo Wilson of Toronto University first defined transform faults to explain the puzzling abrupt endings of such major crustal features as mountains, mid-oceanic ridges, and major faults with horizontal movement. In his now classic work, he suggested that such features were not isolated with abrupt ends but were instead all connected, forming a network of mobile belts within the crust which marked the boundaries of several great rigid plates that fit together to form the earth's surface. He believed that one type of motion which deforms the crust could be changed or transformed into a different type of motion across the junction of two different structures. He called the junction at which a feature such as a fault is changed into another, such as the mid-oceanic ridge, a **transform**. Transform faults have been well verified and are now known to constitute one of the three types of plate boundaries. Along the transform fault boundaries the plate neither gains nor loses surface area during movement.

Mapping of the ocean floor in the eastern Pacific early revealed long faults running east-west which produced cliffs or scarps thousands of feet in height. The faults extend deep into the crust. Such faults have been found on many parts of the ocean floor elsewhere, and some cut across and offset the great Mid-Atlantic Ridge. When magnetic surveys revealed the symmetrically striped

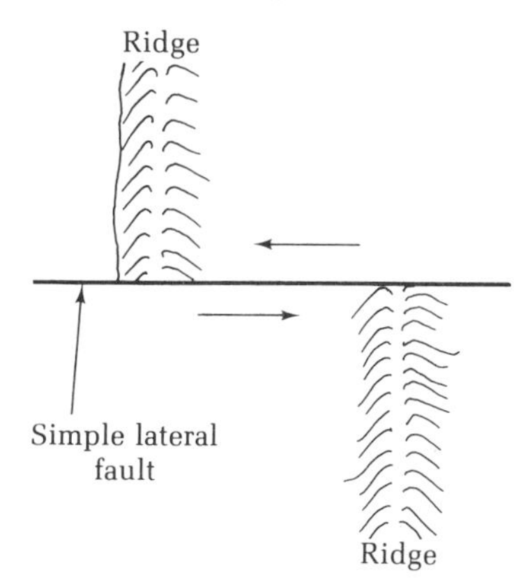

FIGURE 7.17b Two Segments of a Formerly Continuous Ridge Offset by a Simple Lateral Fault. Note that the sense of slip or relative movement shown by the arrows is opposite to that of the transform fault in fig. 7.17a.

pattern of magnetic anomalies on parts of the floor which were also cut by these faults, it was thought that matching of magnetic stripes across the faults would aid in determining fault movements. An unusual feature appeared which seemed to defy explanation. The separation or movement seemed to be left-lateral at one end of certain of the faults and right-lateral at the opposite end (see fig. 7.16, which shows a simple lateral fault). It was this situation that prompted Wilson to his transform fault explanation. He believed that such faults with apparently opposite motion at their ends separated segments of the mobile ocean floor which are spreading at different rates.

Where transform faults cut ridges or trenches, they are not really displaced. The sense of slip along the faults is really the opposite of that which is apparent when one merely looks at the offset of the ridges (see fig. 7.17a). The way in which a transform fault differs from a simple strike-slip fault can be seen in figs. 7.17b, 7.18, and 7.19.

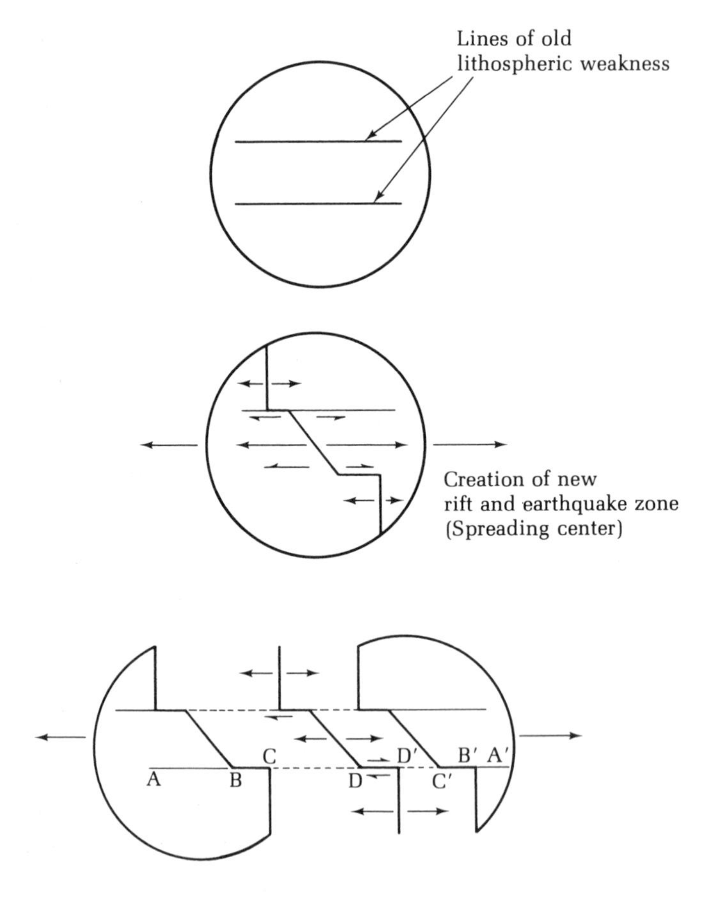

FIGURE 7.18 A Schematic Diagram of the Opening of the Atlantic Ocean Showing That Transform Faults Merely Connect the Spreading Axes and Form Original (Prerifting) Parts of the Plate Boundaries. Segments AB and A′B′ of the east-west fault are older than the spreading. DD′ is the only segment of the transform which is still active. (After Wilson, 1965.)

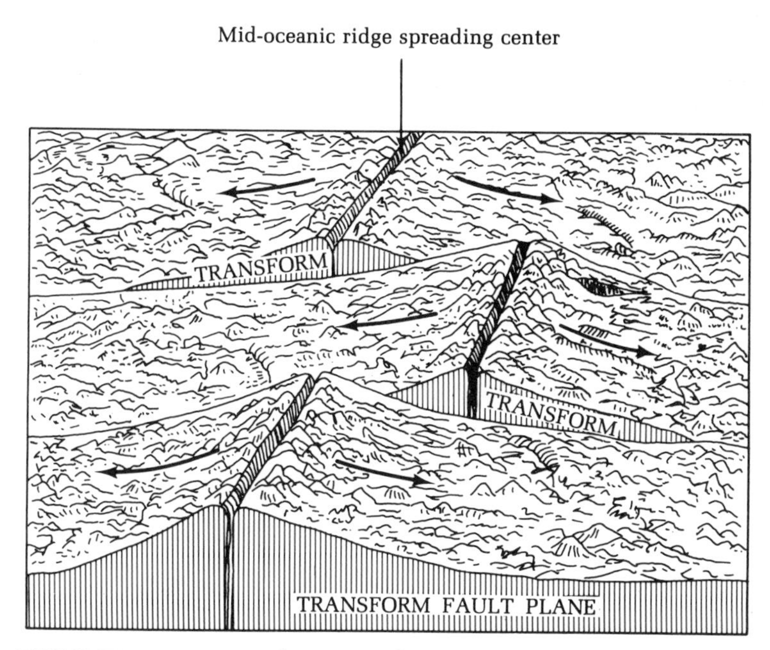

FIGURE 7.19 Transform Faults Connect Ridge Segments Which Are Offset from Each Other. Each offset segment continues its activity of separation with the accompanying formation of new crust by the cooling of upwelling lava along the ridge centers. The transform may also link island arc, deep trench systems with each other or connect a ridge to a trench. (From Mintz, *Historical Geology*.)

These great transform rifts which offset the oceanic ridges and connect the island-arc chains are not what they seem at first sight. When a ridge is offset by a transform fault, the two offset segments of the ridge continue their activity of separation with the formation of new crust by upwelling lava along their centers (see figs. 7.17a and 7.17b). The transform may also link island arc-deep trench systems along which crust dives into the mantle. It was formerly believed that the transforms were zones along which the once continuous mid-oceanic ridges had been broken and the segments moved apart. **It now appears that they really merely connect the spreading axes and form original parts of the plate boundaries.** (See fig. 7.18.) The fracture zones of the transform faults lie at approximately right angles to the spreading center axes and show scarps, and long troughs, and ridges with lines of volcanoes. The height of the scarps along the transform fault depends upon distance from the spreading centers, and it naturally reverses at about midway along its active length since it connects two spreading centers (see fig. 7.19).

The ridges and troughs of the transform fracture zone are hundreds of miles long, five to ten thousand feet high, and tens of miles in width. Although most of the lava is erupted onto the ocean floor along the ocean-ridge spreading centers, some small amount also escapes along the transform fracture zones.

The active part of the transform fracture zone lies between the mid-oceanic-ridge spreading centers which they connect; the transforms are almost inactive beyond the spreading centers. Most transform faults, both extinct and active, cut the ocean floor, but some are found on the continents in Pakistan, New Zealand, and California. The great San Andreas fault of California is now considered a ridge-to-ridge transform about 1000 miles in length along which the western plate moves northward more than 2 inches (5 cm) per year relative to the eastern plate carrying North America. The San Andreas fault connects the Juan de Fuca ridge, southwest of Vancouver Island, with the ridge in the Gulf of California which is a continuation of the East Pacific Rise (see figs. 7.20a and 7.20b).

FIGURE 7.20a Aerial View Northwest along the San Andreas Fault Which Cuts the Continent in Southern California in the area of the Elkhorn Hills and Carrizo Plain. The fault is the boundary along which the lithospheric plates move past each other, averaging about 2 inches (5 cm) per year. The relative motion carries the left-hand plate north, away from the viewer. Major earthquakes occur when stresses build up and friction is suddenly overcome, snapping the two sides into new positions of adjustment. (Photograph by R. E. Wallace and Courtesy of U. S. Geological Survey.)

An underlying assumption in the theory of plate tectonics is that the plates are more or less rigid; this idea seems to be borne out by several lines of evidence, including the very good jigsaw-puzzle fit which can be made between the edges of many of the continents — especially at the 6000-ft-depth level. The magnetically banded rocks of the ocean floor laid out in mirror image fashion to either side of the mid-oceanic ridge centers also support a rigid plate idea. The symmetrical magnetic pattern has not been deformed. The sedimen-

FIGURE 7.20b Aerial View Northwest of the Surface Trace of the San Andreas Fault Zone through the San Francisco Peninsula. The Pacific plate has been sliding past the North American plate along this great transform fault, crushing and grinding the rocks and thus making them more susceptible to erosion. The resulting linear valley was dammed to produce the elongate lakes which serve as a reservoir. (Photograph by R. E. Wallace. The Movement of 1906 — the dashed white line — was approximated by M. G. Bonilla, U. S. Geological Survey.)

tary beds bordering the mid-oceanic ridges are also undisturbed, lying flat in their originally deposited positions. Earthquakes occur along the plate boundaries but not within the plates. All of this evidence suggests that the plates are rigid. (See fig. 7.21.)

If rigid plates form the surface of a sphere, then the motion between any two adjacent plates can be viewed as rotation about an axis which passes through the earth's center. Points near the poles of rotational movement move more slowly and travel a smaller distance than those farther from the pole.

The axis extends from the earth's center to intersect the earth's surface and forms a pole of relative rotation. The relative motion of lithospheric plates may then be viewed as describing circles of rotation. We may therefore speak of minimum movement at the pole and maximum movement at the rotational

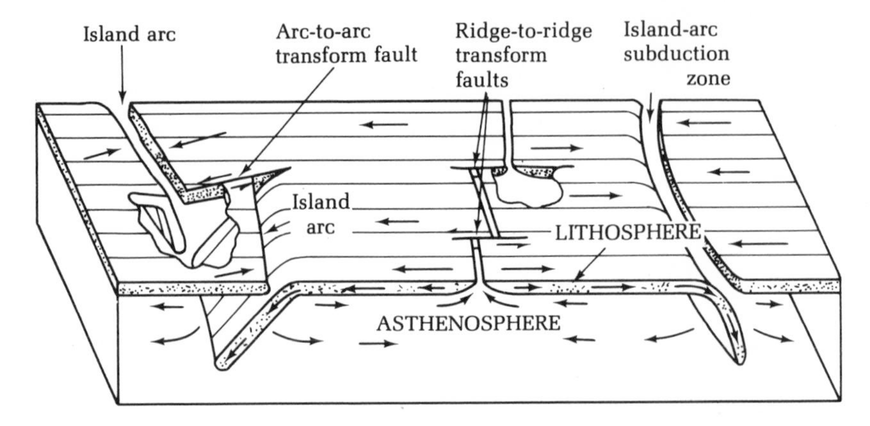

FIGURE 7.21 Schematic Figure Showing Features Which Develop as Rigid Plates of Lithosphere Are Rafted over the Plastic Upper Asthenosphere Layer. Arrows on the lithosphere show the relative motion of adjoining plates. Arrows in the asthenosphere indicate possible flow to compensate for the sinking of parts of the lithosphere. One arc-to-arc transform fault is at the left, connecting two sinking (subduction) zones which are oppositely facing. At the center, two ridge-to-ridge transform faults link segments of a mid-oceanic ridge. A third island-arc subduction zone is at the right. (Reprinted, by permission, from B. Isacks, J. Oliver, and L. R. Sykes, Journal of Geophysical Research 73, p. 5857, 1968, copyright by American Geophysical Union.)

equator. Those plate boundaries which lie parallel to rotation circles are transform faults, along which no plate surface is lost or gained — merely conserved. Correlation of linear magnetic patterns on the sea floor (chap. 8) with the estimated time scale of magnetic reversals on earth permits the calculation of spreading rates for even the oldest ocean-floor rocks. Tracing the position of extinct transform faults allows us to locate ancient spreading poles. (See fig. 7.22.)

Triple Junctions

P. P. McKenzie and W. J. Morgan have recently reviewed the evidence for determining plate movement and have offered new ideas on the geometry of the motion of rigid tectonic plates. Concerned with the "slow evolution of plate boundaries or with changes in their relative motion through geological time," they suggest that the most important long-term changes occur where three plates meet. Such a meeting place they term a **triple junction.** They divide the junctions into stable and unstable types based on whether the junctions keep or change their geometries as the plates drift. The evolution of triple junctions may produce some changes which might have been wrongly interpreted as

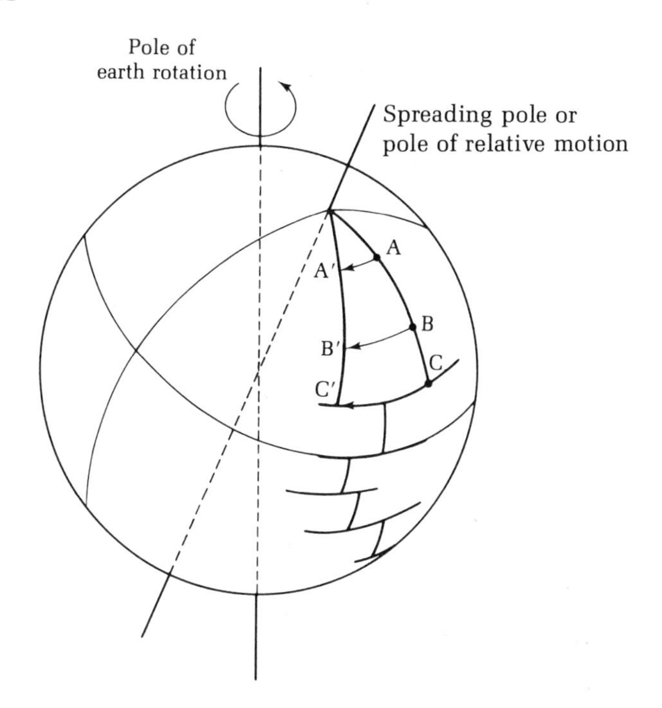

FIGURE 7.22 The Geometry of Lithospheric Plate Movement. An axis or pole of rotation may be drawn so that a set of points on the surface of a sphere can be moved by rigid rotation around the pole. Points *A*, *B*, *C* on a moving rigid lithospheric plate have been carried to new positions *A'*, *B'*, *C'*. Minimum movement occurs at the pole and maximum movement occurs at the rotational equator, 90° away from the pole. The spreading rate along a mid-oceanic-ridge spreading center would obviously be greatest at the equator of relative motion (90° away from the pole of relative motion or, in this case, the spreading pole).

resulting from changes in the direction or amount of relative motion between the plates. The authors present sixteen possible types of triple junctions but note that examples of only six are to be found at present (fig. 7.23). There is clear evidence in the form of magnetic patterns in sea-floor rocks that old triple junctions have been swallowed in subduction zones.

In their discussion, McKenzie and Morgan prefer to define trenches, ridges, and transform faults according to their roles in the creation or destruction of plates; plates are consumed along one side by plunging into trenches; ridges produce new plate material in mirror-image fashion to both of its sides, and transform faults lie parallel to the direction of slip. All three types of plate boundaries actually intersect at the mouth of the Gulf of California (fig. 7.23b). McKenzie and Morgan cite the intersection of the East Pacific Rise and the Galapagos rift zone in the East Pacific near the equator as an example of a three-ridge junction which they consider "the simplest example" with which

Examples of Triple Junctions

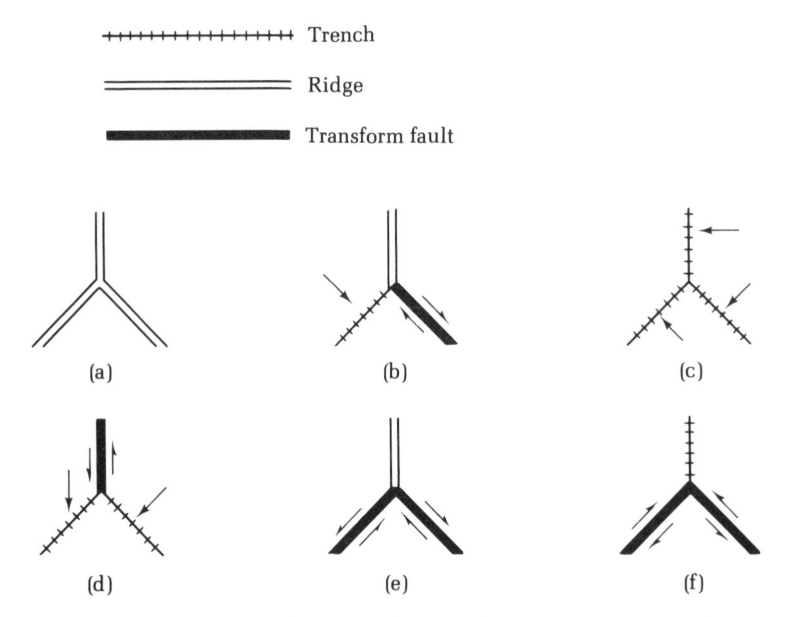

(a) East Pacific Rise and Galapagos rift zone. Also Great Magnetic Bight.
(b) Mouth of the Gulf of California.
(c) Japan trench forks to form Ryuku and Bonin arcs.
(d) Intersection of the Peru-Chile trench and the West Chile ridge.
(e) Owen fracture zone and the Carlsberg ridge. West Chile ridge and the East Pacific Rise.
(f) San Andreas fault and Mendocino fracture zone.

FIGURE 7.23 The Geometry of the Six Types of Triple Junctions Which Are Known to Be Presently Operating. (Reprinted, by permission, from D. P. McKenzie and H. J. Morgan, "Evolution of Triple Junctions." *Nature* 24 (1969).

to work out the general method in determining the conditions which make a junction stable (fig. 7.23a). The Japan trench forks to form the Ryuku and Bonin arcs; here we have an example of a three-trench junction (fig. 7.23c).

The Question of Mountains

The earth's major mountain systems on the continents, such as the Himalayas, Andes, Appalachians, Alps, Urals, and many others, are among the most prominent features of the crust. All have a long and complicated geologic history. Although differing from each other in particular details, they all consist of long, parallel ranges grouped into systems, and they contain mostly sedimentary and metamorphic rocks which are folded and faulted — the result of suffering compression. The central regions of many such mountain systems show huge intrusive granitic rock masses called **batholiths** exposed by erosion of the overlying rocks. Such systems are several times longer than wide and may extend for more than 1500 miles, such as the Appalachians do along the North American east coast.

Some of the sedimentary rocks comprising such mountains show fossils and other indicators such as ripple marks and mud cracks which testify to a shallow-water origin. These shallow-water sediments accumulated to great thicknesses, for the folded rocks of such mountain systems are often 30 to 50,000 feet thick. In some mountain systems a belt made up of these shallow-water shales, sandstones, and limestones is found adjacent to the continent. Other belts contain immature or less weathered sediments which appear to have been deposited around island arcs or deeper ocean water before they were weathered and segregated, thus indicating that the distance from source to site of deposition was small. Some belts show pieces of oceanic crust.

Evidently sedimentary rocks representing diverse environments and fragments of ocean crust have been pushed up or thrust up and over each other in such mountain systems. The thickness of sedimentary rock layers in such systems is greater than that of rocks of the same age located in the adjacent lowlands toward the interior of the continent, and therefore the rocks must have been deposited in a deeper basin.

As early as 1859 James Hall recognized these generalities in mapping the Appalachians; his ideas were extended by James Dwight Dana who coined the term **geosyncline** to describe a huge, long trough in which sediments were deposited over a great interval later to be compressed, folded, faulted, and raised into a mountain range. The geosyncline concept has long been accepted by theorists who regarded continents and ocean basins as unmoving, fixed, and very ancient features of the earth's crust. Plate tectonics brings new interpretations to the history of geosyncline-derived mountain systems.

Perhaps the cool arm of a convection cell turns downward to form a deep elongate basin similar to our modern trenches. Sediments accumulate on the basin floor to great thickness, and their increased mass aids in depressing the floor of the basin. The sediments are lighter than the rocks surrounding the basin and thus have a tendency to rise, but they are held down by the descending convection current. As the sediments sink deeper, they become heated from both depth and radioactivity which turns them into molten granite. The balance between the buoyant tendency of the light basin rocks and the descending convection current may be upset when the convection pattern changes. At that time the molten granite and depressed sediments rise isostatically to great heights, forming mountain systems. Sometimes the process is further complicated by continental collisions in which the "geosynclines" or sedimentary troughs which lay along the margins of both landmasses are sandwiched between them. Such complicated mountain chains resulted when Africa and India crashed into Europe and Asia.

Robert Dietz has recently proposed a model of mountain formation which he thinks applicable to the Appalachians and other mountains formed of sedimentary rocks which were deposited in geosynclines and buckled up along the margins of continents. His model nicely explains the belts of radically different rock types which are pushed against each other in such geosyncline-derived mountain masses (see fig. 7.24).

The theory of plate tectonics holds that mountain chains originate along subduction zones which are marked by volcanic activity, earthquakes, and the

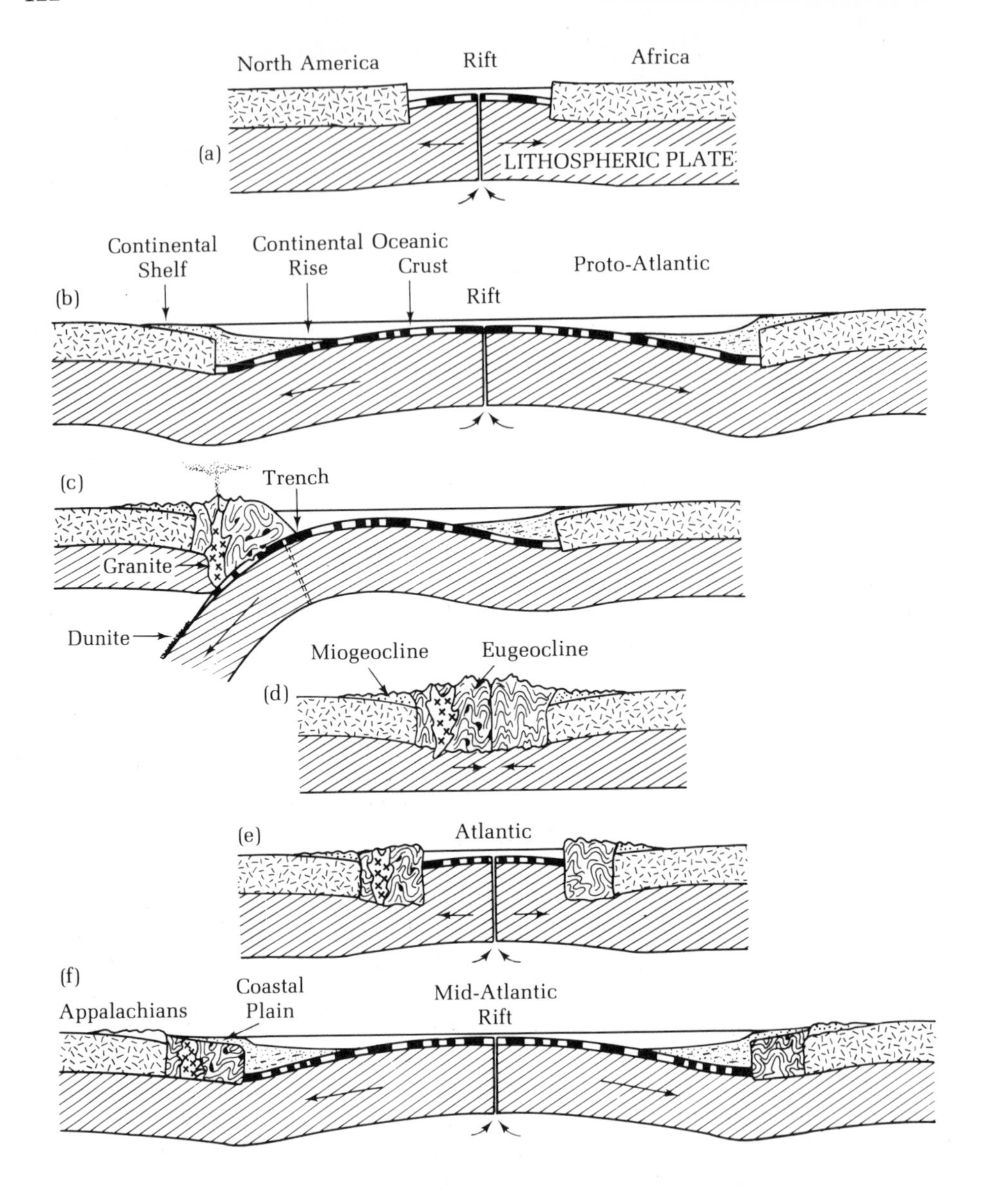

accumulation of sediments which are destined to be plastered onto the continental edge and add to its size. The mountain-building process consists of several stages: (1) lithospheric plates in motion form a subduction or sinking zone which is expressed on the ocean floor as a trench; (2) the underthrust, plunging lithospheric plate melts at depth and produces molten rock which rises to erupt and form an arc-like chain of volcanic islands; (3) the island arc and nearby continent are weathered to yield sediments which are deposited in the adjacent shallow sea. The sediments laid down in the trench and around the island arc are immature, i.e., they have not been weathered and sorted because they are deposited close to their site of derivation. They are mostly **graywackes,** a sort of gray sandstone. The graywackes are often associated with volcanic material presumably derived from the island-arc volcanoes. Closer to

FIGURE 7.24 (Opposite). Stages Which Produce a Folded Mountain Belt. (a) More than 600 million years ago, Africa and North America were torn apart along a rift which acted as a spreading center. Lava welled up to form new ocean floor crust; as it cooled, it recorded the earth's magnetic field for that time; the periodic reversals of the magnetic field were later recorded, producing the striped pattern in the basaltic oceanic crust. (b) The ancestral Atlantic Ocean grew wider as the plates moved apart. At the edge of each continent, sediments accumulated; the shallow-water sand, mud, and limestone collected on the continental shelf to form the characteristic *miogeocline* rocks. In contrast, the deeper-water *eugeocline* rocks deposited on the ocean floor were mostly gray sandstones and volcanic products. Many geosynclines can be divided into these two parts referred to as the *geosynclinal couplet.* The ancestral Atlantic then began to close. (c) The lithosphere cracked and the new plate edge was forced beneath the other, forming a trench. The diving lithospheric tongue was destroyed as it moved deeper into the mantle. As the underthrust plate descended, it buckled the eugeocline, forming the first Appalachians. The eugeocline rocks were then invaded by molten rock which formed deep-seated granite masses and andesitic volcanoes. The Atlantic grew smaller as Africa moved toward North America. (d) The continents collided, forcing their geosynclinal couplets together, folding, faulting, and uplifting them. The line of juncture between the continents was a transform fault along which pieces of mantle rock (ophiolites) were squeezed up. North America appears to have been attached to Africa in this way "between 350 million and 225 million years ago." (e) The modern Atlantic began to open about 180 million yr ago along a new rift line not far from the old one. (f) The Atlantic continues to open today at the rate of a little more than one inch per year near the equator. Several such episodes of spreading and closing with accompanying subduction may have torn continents apart and sutured them together throughout much of earth history, thus producing major mountain belts. (From "Geosynclines, Mountains, and Continent-building," Robert S. Dietz.)

the continent, mature sediments which have traveled great distances collect to form shale, sandstone, and limestone; (4) compression in the subduction zone deforms all of these sediments, but apparently the poorly sorted graywacke sandstone and volcanic material on the ocean side of the volcanic island arcs are folded first, followed by the sandstones, shales, and limestones between the arc and the continent, and finally the rocks of the continental margin are deformed; (5) this entire compressed, greatly thickened rock mass begins to flow plastically; it is pushed to mountainous heights, and its floor sinks deeper into the mantle. Igneous activity at depth begins, and the central region of the compressed mountainous mass may then suffer repeated intrusions of molten material which cools to form igneous rock bodies of various sizes and shapes. The entire region has at this stage been uplifted to form a mountain belt.

Both ancient and more recently formed mountain systems are generally bent in an arc shape away from the old parts of continents. The arc-shaped mountain belts usually show a zone of graywacke (trench deposits) bordered by a belt of volcanic, intrusive, or metamorphic rocks representing the old arc-shaped chain of volcanoes. Still further toward the continent lies a belt of folded and faulted shales, sandstones, and limestones. These three belts are present, with minor modifications, in most of the major mountain systems on earth. The theoretical model of colliding plates of lithosphere seems to explain them nicely.

The possibility of connection between mountain-building activity and changing patterns of convection currents is supported by recently accumulated data which reveal that a short pause in spreading took place about 10 million years ago. Structural studies of mountain systems indicate that there was increased mountain-building activity at that time which resulted in the formation of many new, large structures within the mountains. The models of mountain building offered by plate tectonics do not explain all mountain belts; many geologists think that Precambrian (older than 600 million years) mountain belts may have formed in ways basically different from younger ones because the earth's crust was probably much thinner at that time, thus giving rise to a very different pattern of stresses. A final, all inclusive answer to the mountain-building question still lies in the future.

8

Plates Past, Present, and Future — A Schedule for Drifting

Did the Atlantic Open Twice?

It is believed that the fragmentation of the supercontinent of Pangaea began about 180 million years ago. The present continents were joined together before that time. Earlier we mentioned that very unique kinds of rocks called **ophiolites** are found along narrow zones in certain mountain chains. They consist of masses of very basic igneous rocks such as basalt, gabbro, and olivine-rich rocks which apparently have been thrust upward into overlying rocks as wedges and slivers. Their structure and composition suggest that they may be pieces of uppermost mantle or ocean-floor crust that broke from descending plates and were driven upward into overlying rocks. Ophiolites therefore are believed to lie along the sites of former subduction zones and mark the line along which continents collided to form folded mountain ranges such as the Alps and Himalayas. Since there are mountain belts on earth, including the Appalachians, Urals, and others, which are older than 180 million years and contain ophiolite zones, it suggests that they were formed by collisions at now extinct plate boundaries. Such continental collisions must have destroyed ancient oceans which lay between them and produced such mountain belts with their unique ophiolite zones.

Such an interpretation led Tuzo Wilson to the hypothesis that the Atlantic Ocean has opened, closed, and then reopened to form the present ocean. Sometime during Precambrian time (before 600 million years ago), North America and Africa were torn apart by a spreading rift which produced an ancestral Atlantic Ocean. Sediments accumulated along the margins of the two separating continents for 300 or 400 million years. When currents within the mantle changed their directions, the two continents moved toward each other, resulting in underthrusting of the oceanic plate and creation of the Appalachians (see fig. 7.24). E. A. Hailwood and D. H. Tarling (1973) have shown that paleomagnetic data from Africa, North America, and Europe generally agree with the positioning of a proto-Atlantic in the present site of the North Atlantic Ocean during the lower Paleozoic (600–400 million years ago); thus the paleomagnetic data are consistent with the age of folding of the mountain belts cited by Wilson. The process of crustal deformation which closed Wilson's proto-Atlantic and produced the Appalachians extended from 350 to 225 million years ago. North America and Africa were again torn apart about 180 million years ago along a new rift line which was situated close to the old one. The central part of the North Atlantic Ocean continued to open at the rate of slightly more than one inch per year.

Before turning to a description of the individual plates which make up the lithosphere, we may briefly consider the very basic question about the origin of the supercontinent before any breakup occurred. Patrick Hurley has plotted belts of contemporaneous crustal disturbances and igneous activity on maps which show the northern continents reassembled in their predrift positions. In contrast to Tuzo Wilson, he believes that continental drift has probably been confined to the last 200 or so million years and thus rules out more than one opening of the Atlantic.

Hurley's pattern of age provinces suggests an arrangement into concentric zones, with the oldest rocks at the center and the youngest at the outer boundaries. He believes that such an arrangement supports the hypothesis that Laurasia (the northern supercontinent of some predrift reconstructions) was formed by a process of continental growth in which an original ancient nucleus of crust came to be surrounded by progressively younger crust. The newer crust was derived from the mantle and carried toward and up against the continental nucleus by a moving sea-floor. Apparently the rate of growth of crustal material has increased through geologic time.

Hurley's ideas support the concept that Laurasia continued to grow larger by marginal accretion up until a few hundred million years ago when the process reversed. The present trend of continental breakup and drift then started and has continued until now. He noted that the ancient nuclei or cratons are ". . . not scattered throughout the predrift continental crust as one might expect them to be if they had been initially isolated or separated on the earth's surface and had been brought together by a gathering process prior to the recent drift [within the last 200 million years]. Instead they are all centrally located within the so-called supercontinents of "Laurasia" and "Gondwanaland" and encircled almost entirely by belts of younger continent. This concentric arrangement and the limited degree of scattering of the ancient continental cratons [nuclei] make it difficult to conceive of a series of drift motions prior to the breakup that occurred within the last 200 million years." A decision in favor of Wilson's or Hurley's hypothesis must await further evidence.

The Major Plates of the Lithosphere

The exact boundaries of all of the modern tectonic plates have not yet been definitely located, but most of the boundary areas of the major plates and some of the smaller ones are now well defined (see figs. 7.1 and 8.1). Dietz and Holden summarize plate motions in the following way: "The trench systems, where crust is resorbed [destroyed] may be visualized as an inverted U (the

Pacific trenches) with a stem attached (the east-west Tethys trench). The complimentary system of rifts forms a small circle on the earth within the Antarctic Ocean with two north-south spurs, the Pacific rift system, and the Atlantic rift system. In response to these rift and trench systems, the crustal plates with their included continents have a history of mostly northward and westward displacement. The trenches have tended to remain quite fixed in position, whereas the rifts have migrated so that the plates have changed size. Having no trench or zone of resorption, the Antarctic plate has grown considerably larger; other plates have grown differentially in size and shape, depending on the

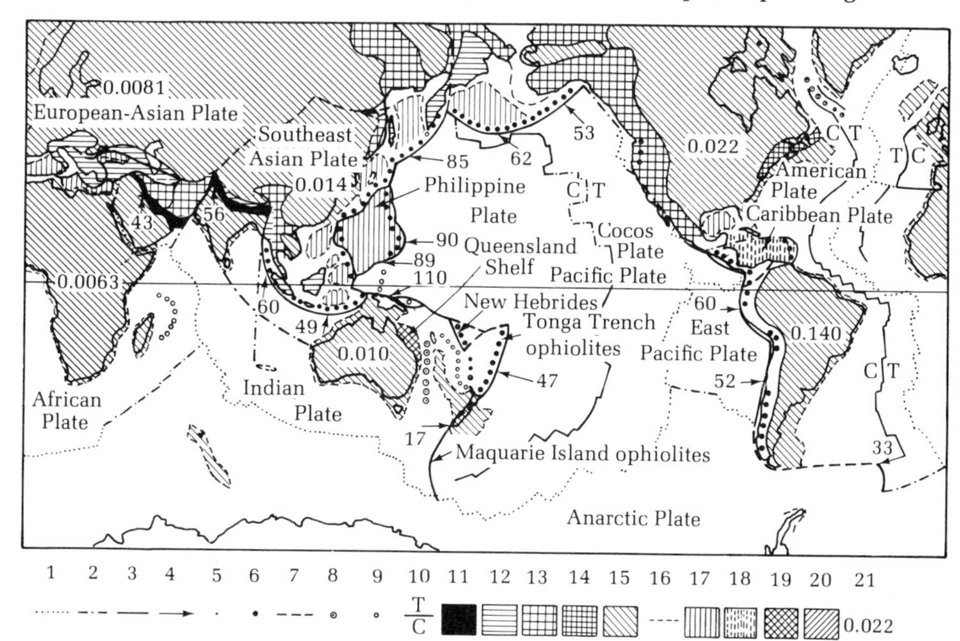

FIGURE 8.1 Tectonic Plates of the Earth (1) Spreading centers, or growing-plate margins. (2) Active transform fault. (3) Trench subduction zone — plate margins are consumed here. (4) Lines showing direction and rate of plate movement expressed in centimeters per year. (5) Localities of diffuse, shallow earthquakes. (6) Extinct, growing-plate margins. (7) Inactive transform fault. (8) Volcanic arc. (9) Extinct volcanic arc. (10) Boundary between Cretaceous (136–65 million yr) and Tertiary (65–3 million yr). (11) Tertiary, collisional, mountain belt. (12) Tertiary folding and metamorphism. (13) Mesozoic (225–65 million yr) mountain belt. (14) Northern Appalachian Mountains. (15) Continental crust. (16) Continental margin. (17) Small ocean basin. (18) Intermediate crust. (19) Extensive limestone rock platform on continental shelf. (20) Thick accumulations of sediment on continental margin at major river mouth. (21) Rate of continental erosion in centimeters per year. (John F. Dewey and John M. Bird, *Journal of Geophysical Research* 75, p. 2627, 1970, copyright by American Geophysical Union.)

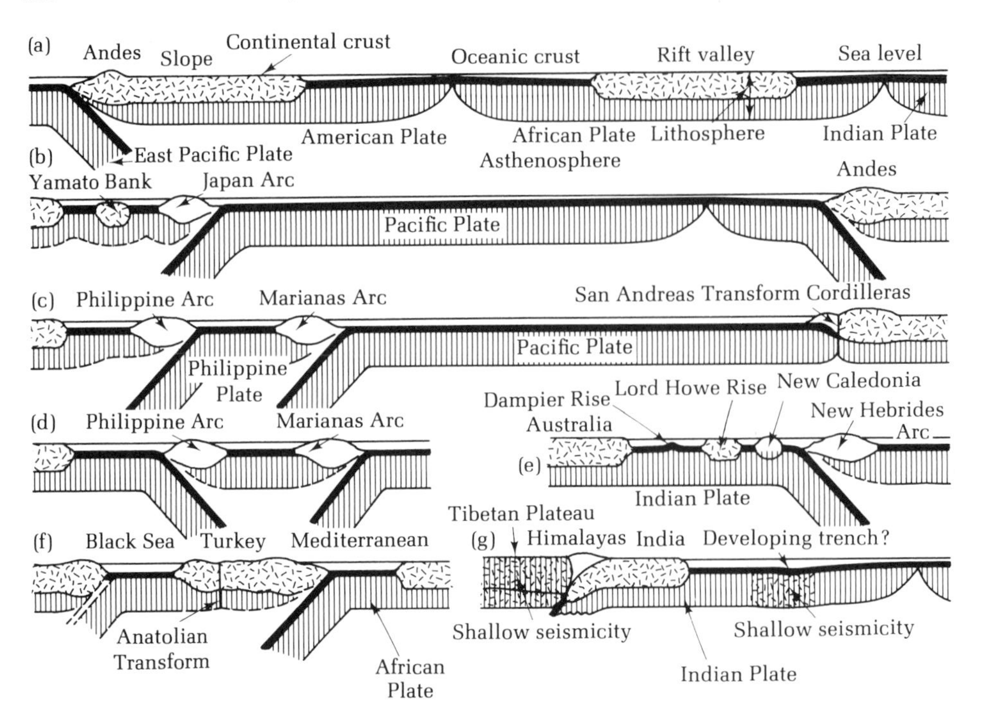

FIGURE 8.2 Schematic Cross Sections Illustrating the Relationships among Plates, Ocean Trenches, Continents, and Mid-oceanic Spreading Centers. Sykes (1970) believes that the belt of shallow quakes reaching from India to Australia represents the beginning of a trench caused by the Indian-Asian collision; before this crash, the Himalaya trench was consuming lithospheric plate but was stopped at collision, and the Himalayas now occupy the old trench site. Because new plate is still being generated in the Indian Ocean, a new trench will probably form along this zone of shallow quakes to consume the plate edge south of the Indian continent. McKenzie believes that when a continent collides with an island arc, a new trench may form on the oceanic side of the arc, thus changing the direction of plate underthrusting. The developing trench shown in (g) is from Sykes (1970). (From John F. Dewey and John M. Bird, *Journal of Geophysical Research* 75, 1970, copyright by American Geophysical Union.)

amount of new crust added with respect to that resorbed. The Pan-Antarctic rift has migrated northward while the Atlantic rift system has migrated westward. Carried like a raft stranded on a moving ice floe, North and South America have moved large distances westward toward the Pacific trench system. Africa, India, and Australia have moved northward as if drawn by the Tethys trench system. Antarctica and Eurasia have remained relatively fixed, although undergoing some rotation." (See figs. 8.2 and 8.3.)

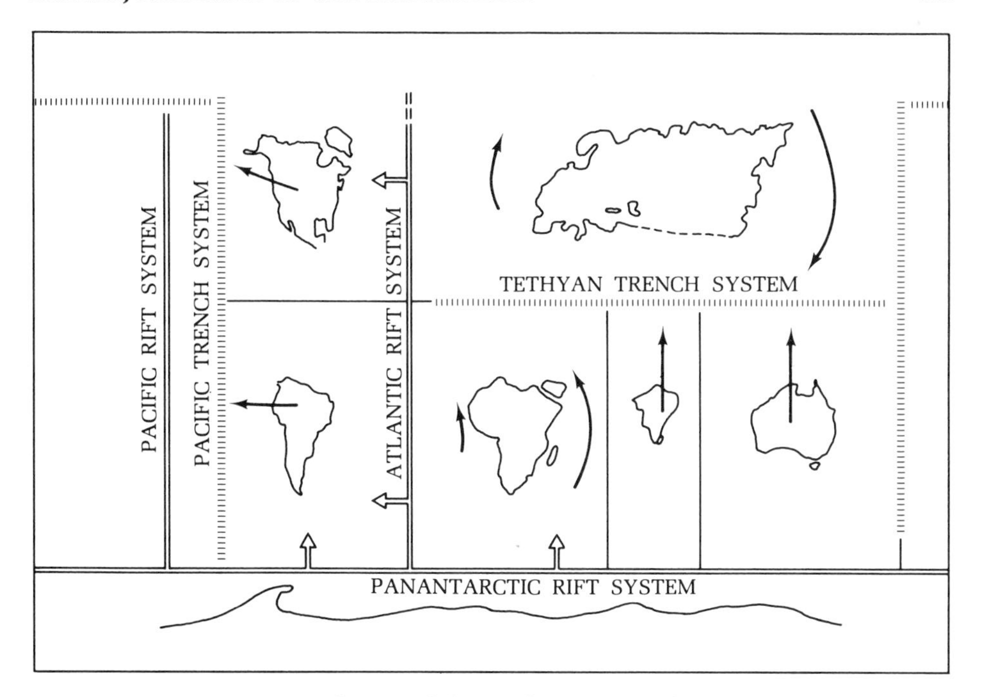

FIGURE 8.3 Greatly Simplified Schematic of the Lithospheric Plates Showing Positions of Mid-oceanic-ridge Spreading Centers (Double Lines), Trenches (Hachured Lines), and Transform faults (Simple Straight Lines). The spreading centers migrate in the directions shown by the double-line arrows. The single-line arrows indicate the past movements of individual plates. (Robert S. Dietz and John C. Holden, *Journal of Geophysical Research* 75, p. 4942, 1970, copyright by American Geophysical Union.)

The eight largest plates (see figs. 7.1 and 8.1) are now identified as follows: (1) the **Eurasian** plate, comprised of Eurasia north of the Himalayas with the possible exception of eastern Siberia which may be on the North American plate. Also included is the Atlantic Ocean east of the Mid-Atlantic Ridge and north of the Mediterranean Ridge; (2) the **Pacific** plate, which carries the Pacific Ocean extending from west of the East Pacific Rise to the deep trenches of the western Pacific, south of the Aleutian trench and north of the Indian-Pacific Ridge; (3) the **African** plate, which includes Africa, the Atlantic Ocean east of the Mid-Atlantic Ridge, north of the Atlantic-Indian Ridge, and south of the Mediterranean region, and the Indian Ocean west of the Mid-Indian Ridge; (4) the **North American** plate, which carries all of North America except southwestern California and Baja (Alaska is questionably included), and the Atlantic Ocean west of the Mid-Atlantic Ridge and north of Central America; (5) the **South American** plate, which includes almost all of South America except a small extreme northern fragment, the Atlantic Ocean west of the Mid-Atlantic Ridge, north of the South Sandwich trench, and south of Central America; (6) the

Antarctic plate, comprised of Antarctica, the Pacific Ocean south of the Indian-Pacific and Juan Fernandez ridges, the Atlantic Ocean south of the South Sandwich trench and Atlantic-Indian ridges, and the Indian Ocean south of the Indian-Pacific and Atlantic-Indian ridges; (7) the **Australian** plate, which includes Australia, the Pacific Ocean west of the New Zealand Ridge and Kermadec-Tonga trenches, south of the southwestern Pacific trenches and north of the Indian-Pacific Ridge, and the Indian Ocean east of the Ninety East Ridge; (8) the **Indian** plate, which carries India and the Indian Ocean east of the Mid-Indian Ridge, west of the Ninety East Ridge and north of the Indian-Pacific Ridge.

A number of smaller plates have been identified, including the following: The **East Pacific** or **Nazca** plate is bounded by the Peru-Chile trench, the East Pacific Rise, the Galapagos Ridge, and the Juan Fernandez Ridge. The small **Cocos** plate abuts the East Pacific plate along its southern edge and the **Caribbean** plate to the north. The **Southeast Asian** plate is marked off by the northern Indian Ocean and western Pacific trenches and the Himalayan and related Asian mountain ranges. The **Philippine Sea** plate is adjacent to the Pacific plate on the east and the southeast Asian plate to the west. Several smaller plates are known along the western margin of Central and North America. Work on delimiting plate boundaries is now in progress and much remains to be done, especially in the inhospitable and poorly known areas of the Arctic and northern Asia.

The Schedule for Continental Departures and Arrivals

In 1912 Wegener portrayed his grand concept of continental reconstruction in maps which have since become classics. He proposed that all of the modern continents were united in a single landmass called **Pangaea** (**pan**, all, **gaea**, earth) from the Greek. In 1937 Alexander Du Toit, in consideration of paleoclimatologic evidence, grouped the southern continents at the South Pole and brought the northern ones together such that the fossil coal forests of the northern continents were astride the equator. Du Toit retained the name **Gondwanaland,** coined earlier by Suess, for the southern supercontinent, but he originated the term **Laurasia** for the separate northern supercontinent.

Later workers made rather elaborate jigsaw fits, and modern versions done by computer techniques show remarkably good coastline matches. However, it was not until 1970 that a reconstruction was accomplished in which the modern continents were moved about in time and space through the use of cartographic coordinates and positioned in absolute rather than relative geographic terms. Robert Dietz and John Holden use a model in which all the continents are united into one (Pangaea), joined not at the present coast but at a depth of about 6000 feet, since this is about halfway down the continental slope and is thought to be about one-half the height of the vertical cliffs formed when the continents were first torn apart. Since the walls thus formed probably slumped to a more gentle, stable slope after separation, the 6000-foot mark is thought to be the position of the original tear. Dietz and Holden were able to show the absolute positions of the continents in their reconstructions by employing a concept first proposed by Tuzo Wilson of Toronto University.

Hot Spots in the Mantle Wilson theorized that there are hot spots or plumes which arise from deep within the mantle and remain fixed in position for 100 million years or more (fig. 8.4). Directly above these stagnant hot spots in the mantle, volcanic openings develop on the ocean floor from which lava is expelled. As the lithospheric plates slip over the hot spots, lava intermittently pours out, forming a line of volcanoes which are older with increasing distance from the hot spot. The bearing of the line of volcanoes indicates the absolute direction of crustal movement past the hot spot.

Such a line of volcanic cones is called a **thread ridge,** or **nematath.** The nemataths extend from the hot spot orifice on the ocean floor outward across both spreading lithospheric plates. The linear island chains of the Pacific often grade from an eruptive volcano at one end to extinct volcanoes, then to wave-cut volcanoes, and finally to coral-atoll-capped ones. Wilson's idea of lithospheric plates drifting across the deep-mantle hot spots seems to explain such a gradational volcanic chain rather nicely. The Walvis-Rio Grande ridges of the southern Atlantic and the Chagos-Laccadive Ridge in the Indian Ocean are nemataths which serve as guides in the positioning of Africa, South America, and India since the end of Jurassic time (about 140 million years ago). A very prominent hot spot under Hawaii has produced that island chain which indicates the direction of motion of the Pacific plate. The dating of volcanic rocks on the different Hawaiian islands provides us with accurate estimates of plate movement past the hot spot. The Pacific plate moved over it northward a few inches per year from about 100 to 60 million years ago, but it changed direction to the northwest and slowed to about 2 inches per year about 40 million years ago. The plate appears to have gained speed about 5 million years ago to a rate of 4 inches per year. It appears that nemataths have produced the lines of basaltic islands in the Pacific which apparently originated far from the mid-oceanic ridges. The Samoan, Society, Tuamoto, and Tubai probably formed from hot spots as did the Hawaiian Islands.

W. J. Morgan has offered recent (1972), additional evidence that the rigid plates are moving over fixed, mantle hot spots. He quantified observations of the parallelism of Pacific Island volcanic chains, noting four locations of present hot-spot, volcanic activity: the Juan de Fuca rise near Cobb Seamount; Hawaii; MacDonald Seamount; and the Pacific-Nazca Rise near Easter Island. In figure 8.5, the four thick lines were constructed by rotating the Pacific plate backward through time over these four hot spots. These four predicted lines agree closely with the direction of the seamount chains in the Gulf of Alaska, the Hawaiian-Emperor chain, the Austral-Gilbert-Marshall chain, and the Taumotu-Line chain. About 25 million years ago the Pacific plate sharply changed its direction of movement, resulting in the bend in all the volcanic chains. Morgan noted that even closer parallelism could be obtained if the hot spots were not considered as absolutely fixed but were instead allowed to migrate about 1/5 inch per year. That is very little as compared to the almost 3 inches per year that the Pacific plate drifts. In figure 8.6 he shows paleomagnetic pole positions based on rocks from Pacific seamounts worked out by Fran-

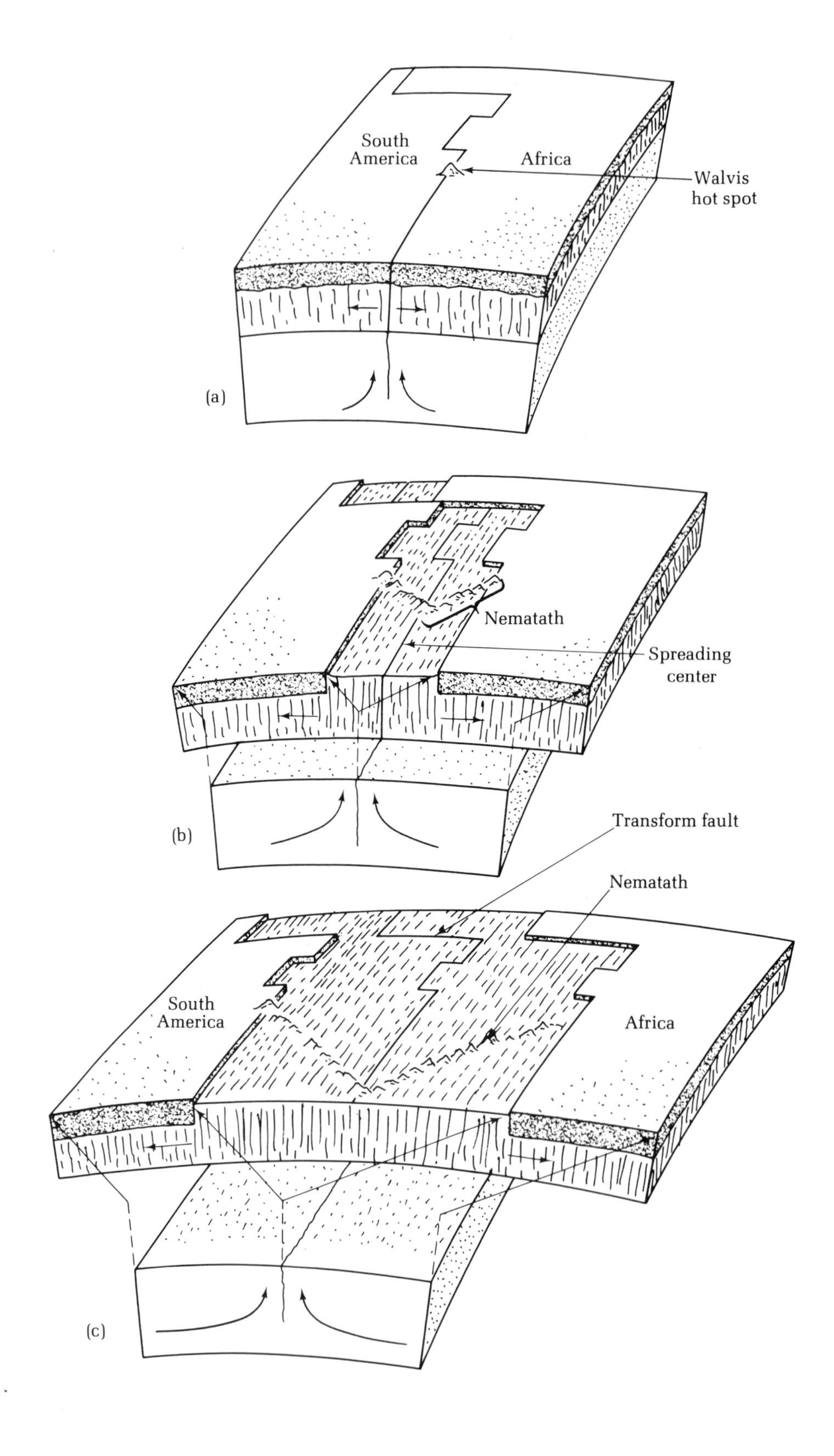
South
America
Africa
Walvis
hot spot
(a)
Nematath
Spreading
center
(b)
Transform fault
Nematath
South
America
Africa
(c)

Figure 8.4 (Opposite). Formation of Nemataths. African and South American plates move apart from a spreading center. The Walvis thermal center or hot spot located on the spreading center has intermittently erupted lava from a deep-mantle source for the past 140 million years. As the plates slide over the hot spot, lines of volcanoes form which trace the absolute direction of plate movement. The oldest volcano is farthest from the hot spot. Such a line of volcanoes is called a *thread ridge* or *nematath*. Note that the two lithospheric plates on opposite sides of the spreading center move relative to each other, while simultaneously both plates may move away from the observer over a hot spot. [Reprinted, by permission, from J. Tuzo Wilson, "Submarine Fracture Zones, Aseismic Ridges and the International Council of Scientific Unions Line; Proposed Margin of the East Pacific Ridge." *Nature*, no. 5000 (1965).]

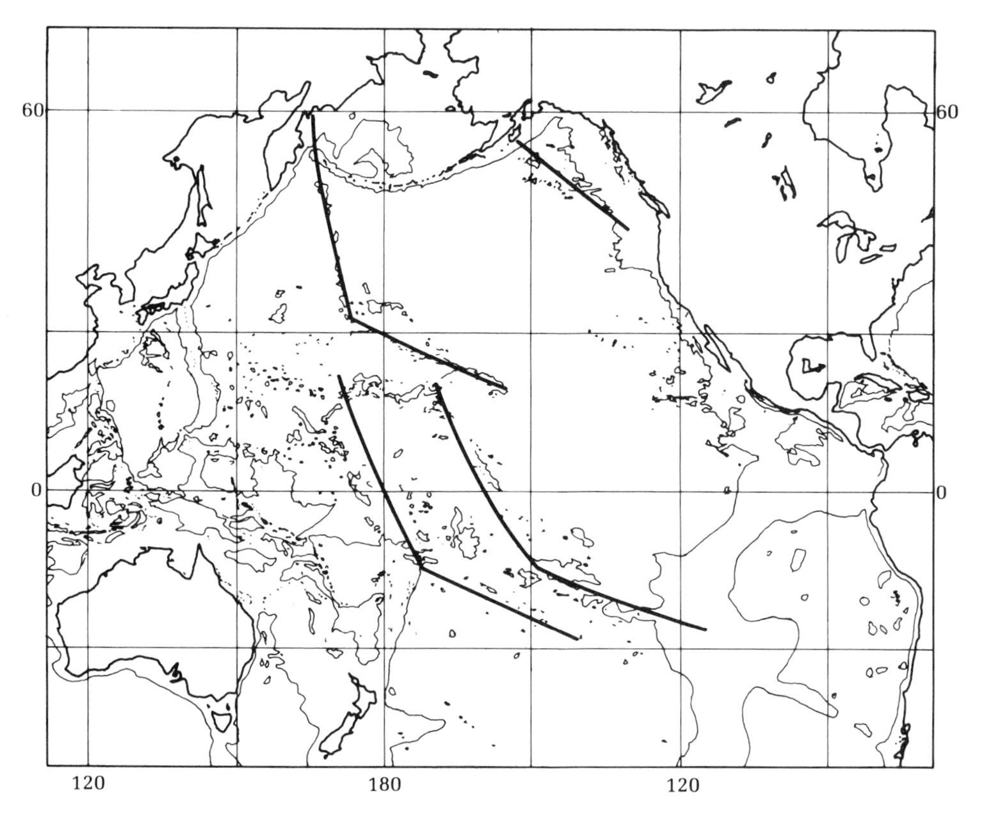

Figure 8.5 Multiple Hot-spot Guides to Plate Movement. The four thick lines were formed by rotating the Pacific plate backward through time over four fixed hot spots. The trends agree closely with those of several different widely separated chains of seamounts (nemataths). (Reprinted, by permission, from W. Jason Morgan, "Deep Mantle Convection Plumes and Plate Motions." *Bulletin of The American Association of Petroleum Geologists* 56 (1972).

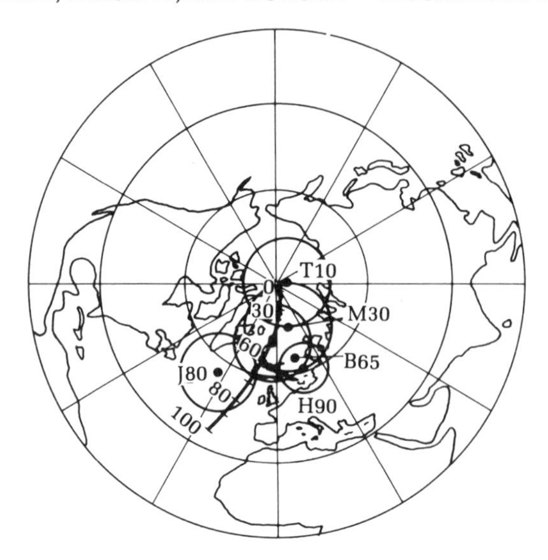

Figure 8.6 Paleomagnetic Pole Positions (Adapted from Francheteau et al., 1970) and Polar-wandering Path Shown by Heavy Line as Predicted by Movement of the Pacific Plate in Fig. 8.5. The numbers indicate pole positions in millions of years ago. The letter shown for each circle identifies Francheteau's seamount group data. (Reprinted, by permission, from W. Jason Morgan, "Deep Mantle Convection Plumes and Plate Motions." *Bulletin of the American Association of Petroleum Geologists* 56 (1972).

cheteau et al. (1970). The Pacific, polar-wandering path predicted by the motion of the Pacific plate (fig. 8.5) is shown by the heavy line. This prediction is based on the assumption that the magnetic pole has not moved relative to the fixed hot spots. In this case the plate movements predicted by nematath trends are supported by paleomagnetic evidence. Morgan shows the plates presently drifting over the fixed hot spots in figure 8.7. The figure is based on relative plate motions as indicated by the trends of faults and spreading rates; to those he added a rotation factor which permits proper rotation of the plate over the hot spots. The lines drawn in fig. 8.7 show direction and are proportional in length to the rate of plate movement. Thus, the figure predicts the trends of island chains and ridges formed on the overriding plate.

Morgan further believes that the stresses on the bottom of plates produced by the material rising in a hot spot are great enough to move the plates and provide the engine for continental drift. He assumes that the hot spots are about 100 miles in diameter and rise about 6 feet per year, originating in the lowest part of the mantle. He cites three main points to support the hypothesis that hot spots drive the continents: (1) most hot spots are located near oceanic-ridge crests or rises, and there is evidence that the hot spots were active before continents split and drifted apart; (2) the gravity readings and surface configuration around the hot spots indicate that the rising columns generate great stresses; (3) estimates of the stresses on the plates generated by the hot

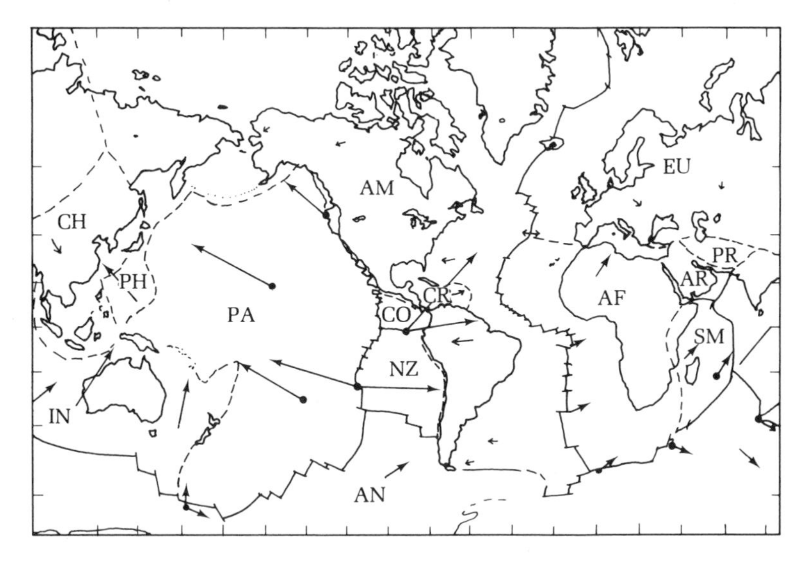

Figure 8.7 Present Movement of the Plates over Hot Spots. The arrow length is proportional to the rate of plate movement. (Reprinted, by permission, from W. Jason Morgan, "Deep Mantle Convection Plumes and Plate Motions." *Bulletin of the American Association of Petroleum Geologists* 56 (1972).

spots seem to be greater than estimates of stresses produced at spreading centers and trenches.

Bernard Minster of the California Institute of Technology believes that various data indicate that "over the past 10 million years the hot spots stayed relatively fixed with respect to each other . . . but we shall have to test it over much longer periods of time." He also noted that relative to the earth's interior, tectonic plates carrying oceans move about five times faster than those bearing continents. Oceanic plates drift an average of 10 cm (4 inches) annually in contrast to continental plates which move an average of only 2 cm (4/5 inch) per year. (See table 7.1) Leon Knopoff of U.C.L.A. holds that if the Antarctic plate may be regarded as more or less fixed, which it seems to be (see fig. 8.16), then it appears that the entire lithospheric shell of the earth may be drifting westward, relative to the earth's interior, at about 1 cm (2/5 inch) per year. Tanya Atwater (1973) noted that Clague and Jarrad (in press) have shown the location of the pole of rotation for the Pacific plate for the late Tertiary which is in agreement with formerly problematic data; thus they support the idea that the earth's spin axis, the magnetic poles, and the Pacific hot spots have been "fixed to one another" during the late Tertiary.

When the information provided by nemataths was considered in conjunction with magnetic anomalies, transform faults, and other evidence, Dietz and Holden were able to reconstruct a series of paleogeographic maps and a schedule of continental departures. Fig 8.8 shows Pangaea as it appeared at the close of the **Permian** (225 million years ago). The Walvis hot spot or thermal center which gave rise to a nematath provided for Dietz and Holden an absolute fixed reference point only from the end of the Jurassic (135 million years ago) to the present. Before this time they estimated the position of West Gondwana back to the Permian by assuming that it drifted only northward prior to the late Jurassic. Dietz and Holden believe that Pangaea (the single supercontinent), surrounded by Panthalassa, the universal ocean, existed at the end of the Permian period about 225 million years ago. Following Tuzo Wilson's idea, they also assume that Laurasia, the northern supercontinent, and Gondwana, the southern, were originally separate but collided in the late Paleozoic era about 300 million years ago to form Pangaea. They parted again along the old collision line during the Triassic period (225–190 million years ago). Although the plates are really moved on the surface of a sphere by rotation about a spreading pole, it is simpler to describe them as moving in a single compass direction as in the following descriptions.

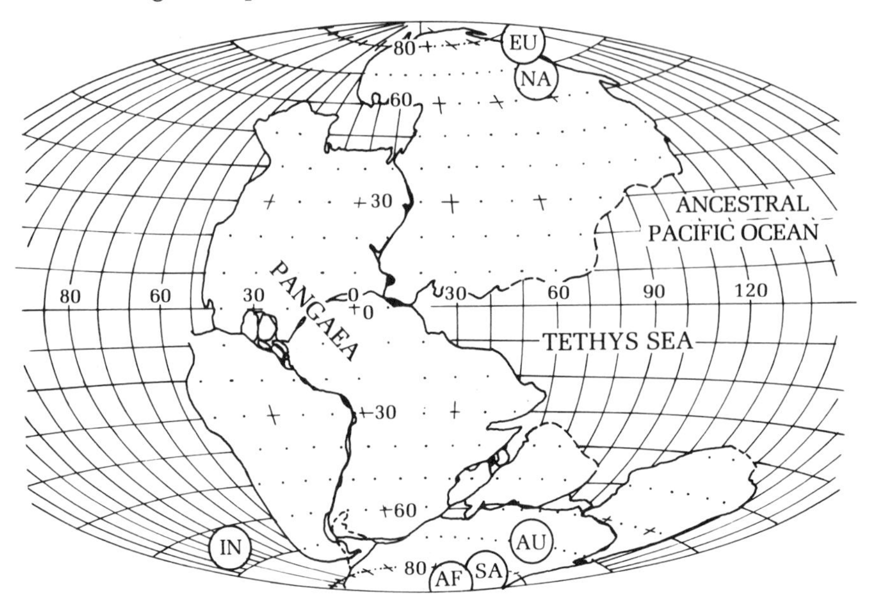

Figure 8.8 All the Continents Assembled to Form Pangaea, the Universal Landmass, As It Existed at the End of the Permian Period 225 Million Yr Ago. The central meridian of longitude is 20°E on this and the four maps which follow, all from Dietz and Holden. Boundaries which have been greatly changed after drift occurred are dashed lines. The solid black areas along continental margins represent large overlaps; the clear areas are gaps in the fit. Certain small land-areas were omitted for purposes of illustration. (Robert S. Dietz and John C. Holden, *Journal of Geophysical Research* 75, p. 4943, 1970, copyright by American Geophysical Union.)

Triassic The middle of the Triassic, about 200 million years ago, marks the beginning of the fragmentation of Pangaea (see fig. 8.9). A rift was opened in the southwest Indian Ocean, moving South America and Africa (west Gond-

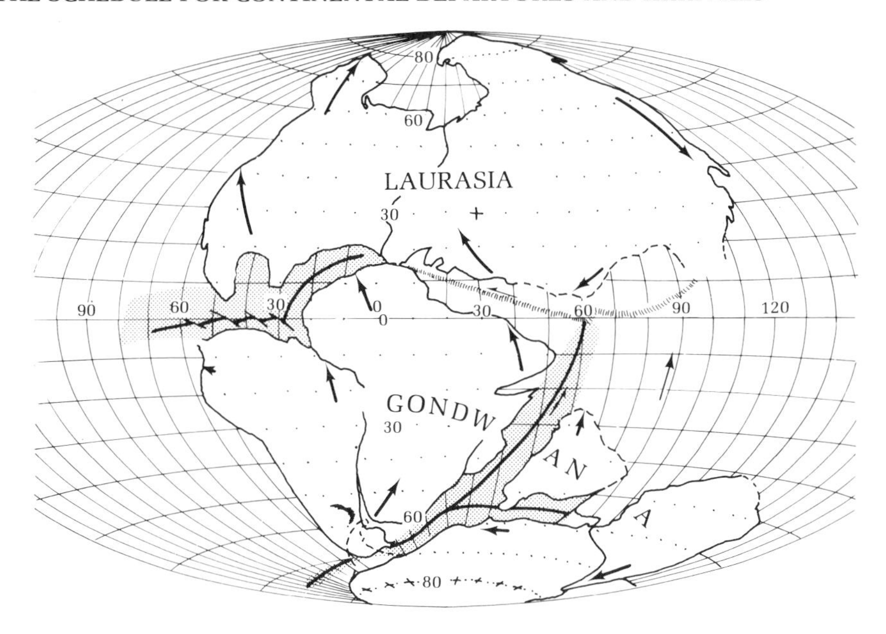

Figure 8.9 Late Triassic World View (180 Million Yr Ago). All workers do not agree on the exact time for the beginning of the breakup of Pangaea; Dietz and Holden tentatively place it at the middle of the Triassic, about 200 million yr ago. The late Triassic view shows the landmasses after 20 million years of drift. Laurasia, the northern group of continents, has broken away from Gondwana, the southern group. India is free and moving north. The South American-African land area has begun to depart from the Antarctica-Australian landmass. The Tethyan trench (hachured black line) extends from Spain to Borneo. Black lines and arrows indicate transform faults. The double-lined arrows indicate direction and amount of drift in proportion to arrow length. The stippled areas represent new oceanic crust created by sea-floor spreading away from mid-oceanic ridges, shown by double lines. (Robert S. Dietz and John C. Holden, *Journal of Geophysical Research* 75, p. 4946, 1970, copyright by American Geophysical Union.)

wana) away from east Gondwana; simultaneously, India was separated from Antarctica along a Y-shaped boundary. North America and Eurasia (Laurasia) were removed from South America and northwestern Africa at the same time along another separate great rift through the northern Atlantic and Caribbean. As continents were torn apart, areas formerly far inland were flooded; revolutionary changes were produced in the crust, seas, climate, and life of massive areas. When the Atlantic opened, great fractures in the crust of New Jersey served as conduits for great outpourings of lava.

Jurassic The Indian and North Atlantic oceans were enlarged further during the Jurassic by northward and westward movement of the sea floor. About 135 million years ago at the close of the period, a rift began, along which Africa was torn from South America. At this time the South Atlantic which lay between them probably resembled the modern Red Sea which is an embryonic ocean. The Atlantic was probably about 600 or 700 miles wide and openly connected to the Pacific. The east coast of the U. S. was oriented east to west at a latitude of 25° north, with water warm enough to support the growth of coral reefs in shallow water along the continental edge. The Bay of Biscay was formed as Spain moved eastward, and the Labrador Sea was opened as Canada and Greenland separated. The ancestral Mediterranean, the great Tethys Sea, narrowed even further in the east. Eurasia moved westward with respect to Africa along a boundary beneath the Tethys Sea. (See fig. 8.10.)

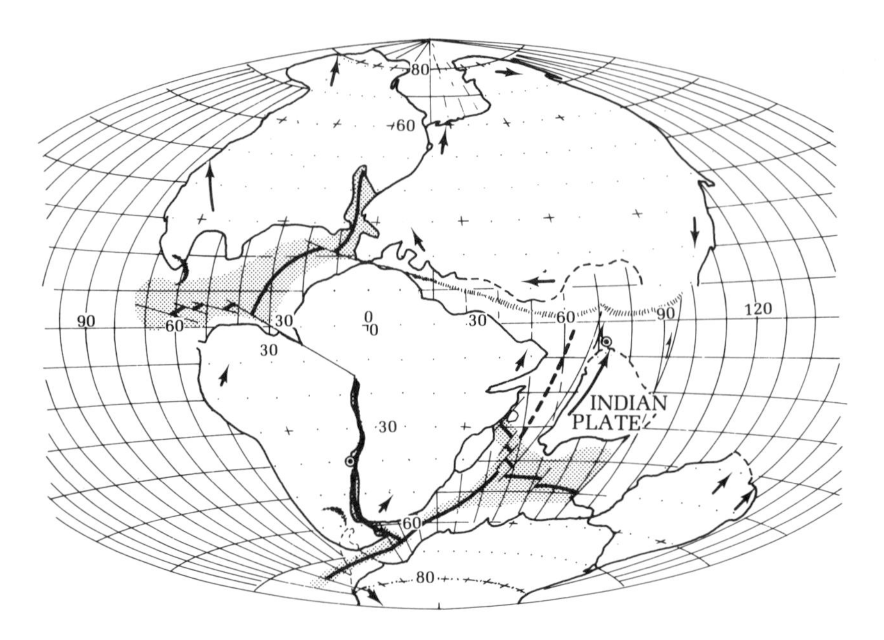

FIGURE 8.10 A View at the End of the Jurassic Period, 135 Million Years Ago, Showing the Birth of the South Atlantic Ocean as Africa and South America Moved Apart. The North Atlantic and Indian oceans had opened much further, and the Tethys Sea began to narrow at its eastern end due to the clockwise rotation of Eurasia. The Indian plate was moving toward and was destined to pass over a hot spot (shown by a black dot in a circle) which erupted basalts, forming the Deccan Plateau. The hot spot continued its outpouring to form the Chagos-Laccadive Ridge, trending north-south in the Indian Ocean. In the latest Jurassic, the Walvis hot spot formed which later provided a fixed reference point used in measuring drift and locating continents in absolute geographic terms. (Reprinted, by permission, from Robert S. Deitz and John C. Holden, *Journal of Geophysical Research* 75, p. 4947, 1970, copyright by American Geophysical Union.)

Cretaceous At the end of the Cretaceous period (65 million years ago) the South Atlantic Ocean was almost 2000 miles wide, and the North Atlantic was almost half that width. The Caribbean region had come under compression rather than tension, and Africa had moved northward 10°, rotating counterclockwise against the clockwise-drifting Eurasian plate, resulting in the further narrowing of the once great Tethys Sea. By this time the modern continents were recognizable, except for Australia, which was still connected to Antarctica, and eastern Greenland, which was contiguous with northern Europe (see fig. 8.11). The North and South American plates drifted rapidly westward and must have been subducted into a now extinct north-south system of trenches.

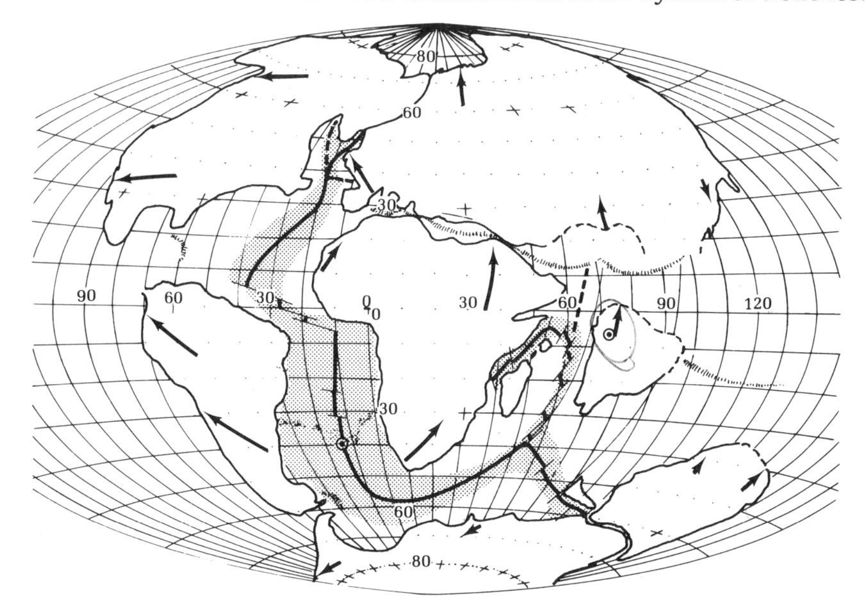

FIGURE 8.11 The End of the Cretaceous Period, 65 Million Yr Ago. The earth after 135 million years of drift appeared close to its modern configuration. The South Atlantic opened greatly and the Mediterranean Sea developed its almost modern form. The once great Tethys Sea became nearly completely closed. The nemataths on the South Atlantic Ocean floor, produced by the departure of the African and South American plates from the Walvis hot spot, were hundreds of miles long. Australia remained attached to Antarctica. A great north-south trending trench probably was situated in the Pacific to subduct the leading edges of the westward drifting North and South American plates. (Robert S. Dietz and John C. Holden, *Journal of Geophysical Research* 75, p. 4951, 1970, copyright by American Geophysical Union.)

About 100 million years ago North America ran into this trench, resulting in the formation of a mountain range along the site of the present California coast ranges. The trench was destroyed as North America rode over it because the lighter, more buoyant rock of the granitic continental crust does not get sucked deep into the trenches as does the heavier oceanic crust.

During the Cretaceous period South America ran into the Andes trench, but instead of overriding it, as in the case of North America, it pushed the trench westward, during which the great Andean belt of folded mountains was built. Dietz and Holden believe that the Andes trench originally dipped westward but was "flipped over to its present eastward dip."

India, which probably separated from Antarctica-Australia during the Triassic, had drifted almost 2000 miles away before the Cretaceous period ended 65 million years ago. Madagascar split from Africa along a branch of the great Carlsberg Rift and moved away as a separate subcontinent.

Cenozoic The continents are shown in their modern positions in fig. 8.12. New ocean floor produced by sea-floor spreading during the past 65 million years of Cenozoic (65 million years ago to the present) time is represented by the shaded area; this area covers about one-half of the entire ocean floor.

The Mid-Atlantic Rift probably opened into the Arctic during the early Cenozoic, severing Greenland from Europe. New Zealand separated from eastern Australia. North and South America were reconnected due to volcanism and crustal movements which elevated the Panamanian neck between them.

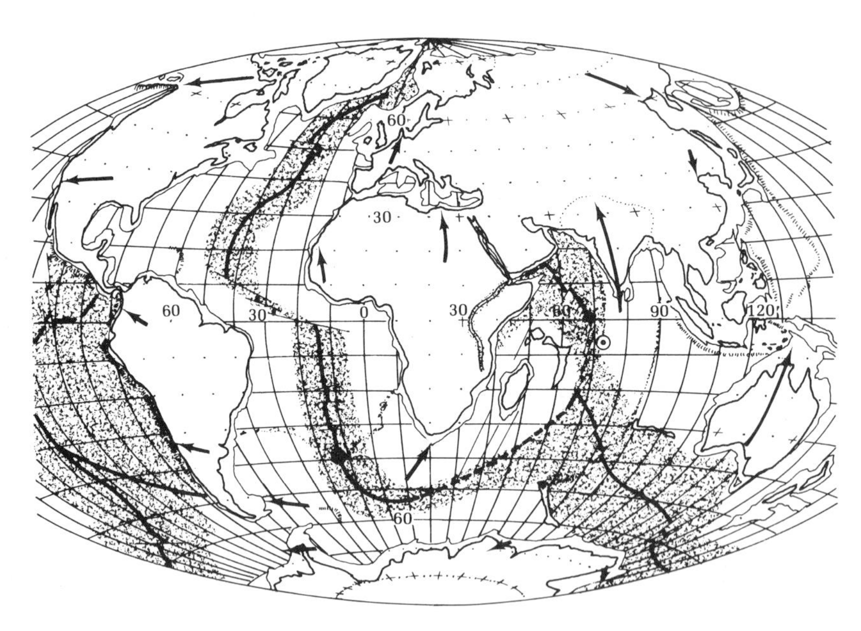

FIGURE 8.12 The Continents in Their Present Position with Sea Floor Created during the Past 65 Million Yr Shown as Stippled Area. About one-half of the ocean floor was renewed by spreading and subduction during this time. India has crashed into Asia to produce the Himalayas; Australia has split from and moved far north of Antarctica. (From Robert S. Dietz and John C. Holden, *Journal of Geophysical Research* 75, p. 4954, 1970, copyright by American Geophysical Union.)

The northern edge of India collided with the southern edge of Asia and was thrust below it, pushing up the folded Himalaya Mountains.

The Australian plate separated from Antarctica about 50 million years ago. It is now drifting northward about 3⅓ inches (8 cm) per year.

Late in the Cenozoic, Arabia was split from Africa along an arm of the Indian Ocean rift. This gave rise to the Gulf of Aden and the Red Sea, both of which are probably very youthful oceans. The floors of both lack continental rocks and appear to be typically oceanic in character. Their axes seem to be spreading centers which are offset in places by fracture zones. The rocks in each exhibit the same pattern of magnetization found elsewhere in the world on the flanks of spreading centers, but since they are embryonic, they contain only the most recently formed parts of the pattern. Separation of both started in the Miocene epoch about 20 million years ago. This opening involved slippage of the crust along the rift of the Jordan Valley with about 50-mile northerly displacement of the eastern block relative to the western side. It also appears that the east African rift valley opened by about 40 miles. See figures 8.13, 8.14, 8.15, and 8.16 for three-dimensional views of plate movement.

Future Dietz and Holden have speculated on what the earth may look like 50 million years from now (see fig. 8.17). The Atlantic and Indian oceans will exand further while the Pacific will be reduced. Australia will drift northward up against the Eurasian plate. The Mediterranean Sea will be crushed by the northward movement of Africa, which also will close the Bay of Biscay; simultaneously, the eastern part of Africa will split off along the great rift presently traced by the great African lake chain. Crustal compression will raise new land in the Caribbean. The Baja Peninsula and a slice of California on the west side of the San Andreas fault will be cut off from the mainland and drift to the northwest such that in about 10 million years, Los Angeles will be carried offshore opposite San Francisco. Los Angeles will be destroyed in about 60 million years when it is pulled down into the Aleutian trench. The continuing expansion of the Atlantic and Indian oceans will reduce the Pacific in size, while the Antarctic plate may rotate slightly in a clockwise direction. In spite of the fact that the Pacific Ocean spreading centers are opening at the fastest rates (see table 7.1), the area of the Pacific is being reduced because most of the active trenches at which crust is destroyed are located in it. If the present activity is projected into the future, perhaps 50 or 60 million years, the west coasts of North and South America will collide with the Philippines and Japan. W. J. Morgan hypothesizes that a hot spot or deep-mantle convection plume is located beneath the region of the Snake River basalts; he suggests that this hot spot which he terms the **Yellowstone plume** "foretells" the breakup of North America.

An understanding of plate movement, both past and present, is a powerful interpretive tool which may allow us to make predictions and retrodictions about many phenomena which have long defied explanation. Earthquake prediction is not yet precise enough to foretell hours, days, or even months, but we now know that strain accumulates if small movements do not take place periodically along plate boundaries. If the fault has been locked by friction for a long time, it may be released all at once in a giant slippage to produce major earthquakes. It may be possible in the future to control earthquakes by helping

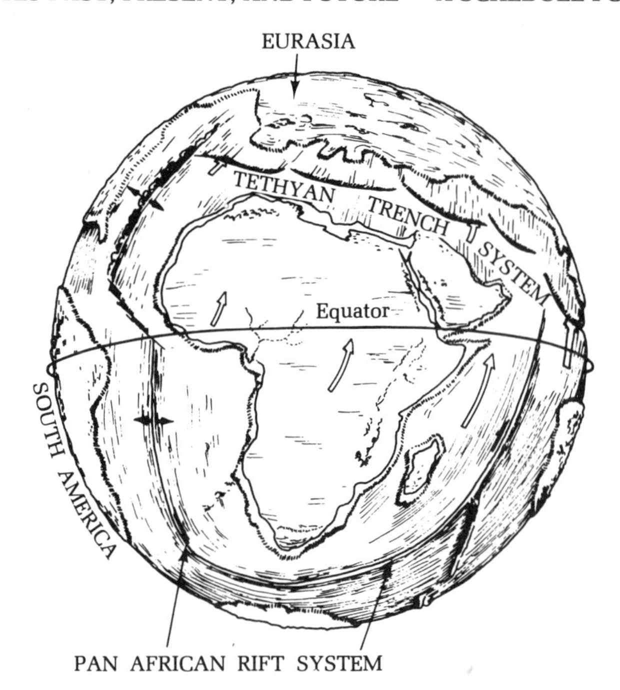

FIGURE 8.13 The African Plate Has Rotated Counterclockwise While Drifting Northward. This movement has narrowed the formerly great Tethys Sea, producing the modern Mediterranean as its vestige. The northern edge of the African plate has been plunging into the mantle along the Tethyan trench. The Mid-Atlantic Ridge has been a spreading center from which the African and American plates still depart. (Reprinted, by permission, from Robert S. Dietz, "Plate Tectonics, Sea-floor Spreading and Continental Drift." *Journal of College Science Teaching,* October 1972.)

gentle slippage to take place by pumping water into the fault zone to lubricate it or perhaps even triggering a number of small harmless quakes by nuclear explosions and thus to relieve stress.

Tectonics and Mineral Resources

The configuration of continents and oceans and the major structural features of the earth's crust are controlled by plate movement; we might also expect that the distribution of the earth's natural resources depend on plate tectonics.

The metallic ores of the earth are found in regions which have been invaded by molten rock. As the intrusive molten rock mass cooled, various ores formed at different levels containing gold, lead, zinc, copper, etc. When these ore bodies are subsequently eroded, the various minerals are sorted and deposited by stream action, sometimes producing valuable secondary or placer ore concentrations. The primary ores formed from molten intrusions are controlled by movement of the plates. Recently, Fred Sawkins of the University of Minnesota has noted that the majority of metal-ore deposits called **sulfides** (metal atoms

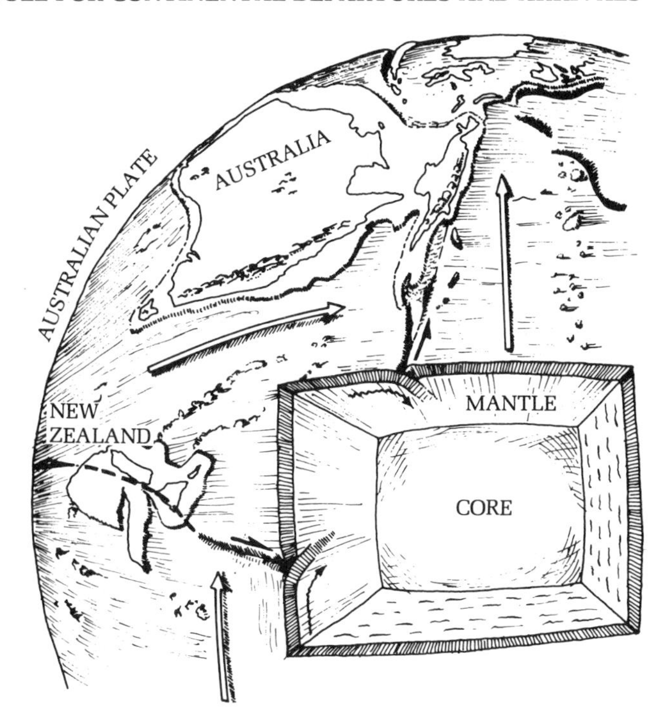

FIGURE 8.14 The Australian Plate Broke Away from Antarctica about 50 Million Years Ago in the Eocene Epoch. It is presently drifting northward about 3⅓ inches (8 centimeters) per year. The Australian plate is moving at almost right angles to the Pacific plate; therefore the boundary between them is a subduction or sinking zone for the Australian plate and at the same time is a transform fault zone for the Pacific plate. The eastern edge of the Australian plate, which runs through New Zealand, is just the reverse of the plate's northern edge. (Reprinted, by permission, from Robert S. Dietz, "Plate Tectonics, Sea-floor Spreading and Continental Drift." *Journal of College Science Teaching*, October 1972.)

united with sulfur) are distributed along old and new plate boundaries at which an oceanic plate dives beneath the edge of a continent. It is thought that the solutions rich in minerals are derived by melting of the sinking plate edge and precipitate along such convergent plate boundaries. The great sulfide ore deposits of the Coast Ranges, Rockies, Andes, Philippines, Japan, and others, and the sulfide-related gold deposits of Alaska, Canada, California, Brazil, South Africa, and Australia are all situated at ancient boundaries along which plates collided.

Divergence at plate boundaries also appears to generate rich metallic-sulfide deposits. Three small depressions in the central part of the Red Sea, which represents the earliest stage in the growth of an ocean by spreading plates, are filled by sediments impregnated with sulfide minerals. These sediments are rich in iron, zinc, copper, and gold which probably arose from hot-water solutions from volcanic sources along the spreading center.

Although thorough systematic sampling has not yet been done, some evi-

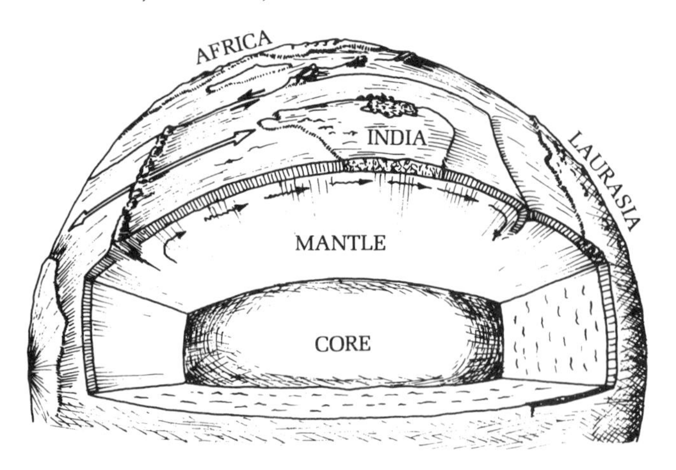

FIGURE 8.15 Since Breaking Away from Anarctica in the Triassic, India Has Rapidly Drifted Northward, Colliding with Asia about 20 Million Years Ago. The Himalayan and related mountain chains were thrust up as a result. The very great thickness of continental type crust (sialic) under the Himalayan region is believed to be due to "underthrusting and dovetailing" which produced a crustal slab "two continents" thick. (Reprinted, by permission, from Robert S. Dietz, "Plate Tectonics, Sea-floor Spreading and Continental Drift." *Journal of College Science Teaching,* October, 1972.)

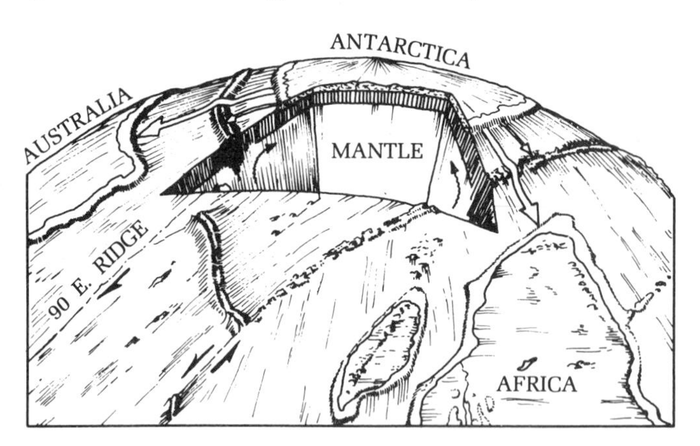

FIGURE 8.16 The Antarctic Plate Appears to Be the Most Stationary of All the Plates Because It Is Surrounded by a System of Transform Faults and Ridges but Lacks Trenches. It has grown larger by the addition of crust along the ridge spreading centers. The departure of Africa and Australia from Antarctica and their drift northward has been accomplished by migration of the mid-oceanic ridges. The ridges have moved in the same direction as the drifting continents but at only half the continental velocity. (Reprinted, by permission, from Robert S. Dietz, "Plate Tectonics, Sea-floor Spreading and Continental Drift." *Journal of College Science Teaching,* October 1972.)

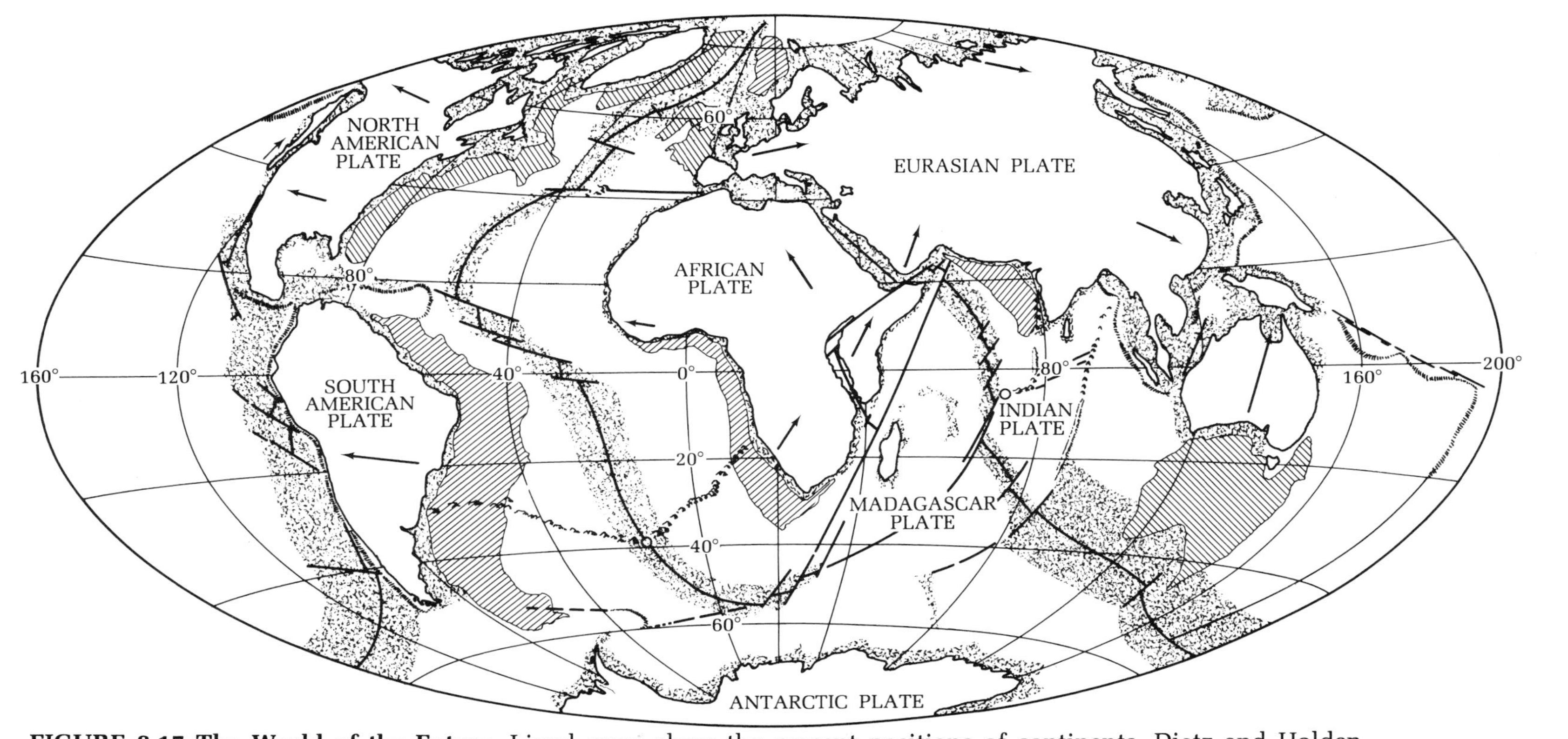

FIGURE 8.17 The World of the Future. Lined areas show the present positions of continents. Dietz and Holden envision such a configuration from extension of present-day plate movements 50 million years into the future. The Atlantic and Indian oceans will expand greatly, reducing the Pacific in size. Australia will drift northward to collide with the Eurasian plate. Baja California and a slice of California to the west of the San Andreas fault will split from the continent and drift to the northwest; this movement will bring Los Angeles offshore from San Francisco in about 10 million years. Southeastern Africa will split off and move northeast to form a separate plate. (From "The Breakup of Pangaea," Robert S. Dietz and John C. Holden. Copyright © October 1970 by *Scientific American*, Inc. All rights reserved.)

dence suggests that volcanic hot-water processes along the major mid-oceanic-ridge spreading centers are concentrating metals in the sediments there. Such sediments are usually enriched in iron, manganese, copper, nickel, lead, and other metals. Such modern zones of mineral formation have their ancient counterparts in the zinc, lead, and silver deposits of middle Paleozoic age found in Newfoundland and Ireland. Here the great rift which opened to form these deposits later widened and the continents drifted apart, forming the Atlantic Ocean. The continuation of the mineral-rich zone should be traceable into the sea off the edges of both continents. In 1963, Gass and Masson-Smith proffered the idea that large, igneous-ore bodies containing copper and nickel found in Cypress, Cuba, Turkey, Greece, New Guinea, and elsewhere, have been pushed up over the land from their origin on the ocean floor. Such a new interpretation may facilitate different approaches to prospecting. It has long been known that some fault zones are the site of great mineral development; in contrast, other major faults such as the Great Glen in Scotland show no mineralization. Plate tectonics may provide answers to such questions and help in delimiting areas for new prospecting.

Petroleum forms from accumulations of plant and animal remains which are trapped in environments which are deficient in oxygen and tend not to support life. This allows the preservation of the organic matter. Such deep-sea environments are created where lithospheric plate edges dive beneath the margins of continents to produce deep-sea trenches. Both the trenches and their associated volcanic-island-arc chains produce areas suitable for the formation of petroleum. Organic material is swept into and trapped in the trenches and island-arc areas; such areas have a configuration which naturally restricts oceanic circulation, thus reducing oxygen and favoring preservation of organic matter. These areas later undergo tectonic deformation of the organic-rich sedimentary beds to produce geological structures which serve to trap and store petroleum formed from the organic material.

It appears that environments suitable to the formation of petroleum may also form at divergent plate boundaries that open under continents. Here, as the pieces of the continent are rafted apart, a sea with restricted circulation and low oxygen forms between them. If the sea water evaporates faster than streams and rainfall replace it, salt deposits may form. The beds of organic matter and salt become deeply buried. The organic material later gives rise to petroleum. The rock-salt beds, because they are lighter than the surrounding rocks, rise to form dome-shaped structures which serve to trap the oil. The Red Sea appears to be such a divergent boundary area; here beds of rock salt and mud rich in organic matter more than 3 miles thick have been found.

Continental movements may help us to understand conditions of sedimentation which give rise to a sequence of beds grading from oceanic to continental to evaporites. It has been suggested by Belmonte el al. (1965) that the oil-bearing beds and salt domes of Gabon in west central, coastal Africa and their counterpart in eastern Brazil originated at the time that the sea flooded the area between the two continents as they first drifted apart in Mid-Cretaceous time about 100 million years ago. The diamond deposits of west Africa are comparable to those of northeastern South America because they, too, were once connected.

The formation of many deposits are controlled closely by climate or by the plants and animals that flourish only in specific climates. Various types of salt deposits, oil, and ores of certain minerals, such as aluminum, all form in great concentration only in hot climates. The determination of ancient pole positions and latitudes which determine climate would aid greatly in exploring for such deposits.

Some Objections to Plate Tectonic Theory

A number of workers have pointed to problems which do not appear to be explained by plate tectonics.

(1) The jigsaw fit of the continents appears good in most places, especially in view of the long period of separation involved, but all reconstructions are not able to fit in Central America because the fits with other continents require that North and South America be too close together. Spain presents something of a problem in that it must be rotated independently of Europe to fit into the Bay of Biscay.

(2) Trenches, bordered by their volcanic island arcs, are supposed to be the sites at which the plate edges turn downward into the mantle; thus we should expect the sediments of such basins to be deformed. Instead, they are horizontal. Since the methods used in such determinations have only probed the upper sediments of the sea floor, it may be that the deeper beds have been disturbed by subduction.

(3) Movement along the transform fault zones has been verified by earthquake waves, but these regions also show undeformed sediments that suggest either very slow movement or none at all.

(4) Elongate troughs of sedimentation (geosynclines) which formerly occupied the sites of and gave rise to the great mountain systems are explainable if they are located along the continental margins. Problematically, there are a number of such apparently geosyncline-derived mountain systems which are situated in the interior of the continents. The Mesozoic-age Rockies of North America are thus difficult to explain. James Gilluly (1971) has pointed out importantly that plate tectonics adequately explains the deformation of some areas such as the Alps, western North America, and the Himalayas, but does not satisfactorily deal with the interior mountain ranges of Asia, the Caucasus, and the eastern Cordillera [Rockies] of North America.

(5) According to plate tectonics, the horizontal motions of the lithosphere are the principal crustal-deforming movements of the earth. Therefore, they may be transformed into vertical or inclined movements of crust only where plates come together. Thus, the vertical and inclined movements in the lithosphere are not considered by this hypothesis to be primary, independent motions, but merely movements which result from horizontal plate movements. Some workers believe that primary, independent, vertical movements of the lithosphere take place, resulting in the vertical uplift or depression of whole continents or major parts of continents. They also believe that such movements directly reflect the vertical forces arising from physiochemical processes operating in the mantle below.

(6) Considering the plates as "rigid blocks" which can be deformed only along their boundaries appears questionable. It seems in disagreeement with the existence of large-scale, intense deformations including severe folding and faulting frequently accompanied or followed by intrusions of molten rock which occur in the interior of continents. Some of these severely deformed areas have never been located at the margins of continents.

(7) Although plate tectonics offers only spreading plates as the mechanism by which to produce oceans and form new oceanic crust, it appears that geologically recent ocean crust has formed in such places as the Black Sea and western Mediterranean.

These areas lack the ridges with tensional rifts along which ocean crust is generated elsewhere.

(8) Many workers have raised doubts about some aspects of almost all lines of evidence in support of drift and plate tectonics. Paleomagnetic data are not always consistent, nor do they always coincide with paleoclimatic and other kinds of evidence. Serious exceptions could be cited for all of the generalizations in support of drift which have been discussed, but it appears that plate tectonics answers more questions than it raises.

The evidence for acceptance of the major components of plate tectonics theory is very good but not absolutely conclusive. A more definitive judgment must await further evidence as this geologic revolution unfolds.

9

Historical Summary

The idea of drifting continents has met great opposition from much of the scientific world; it has been considered favorably in North America only recently. On close examination, we find that the story of continental drift bears out the philosopher's charge of narrow scholasticism and close-mindedness against the scientific community, an apparently valid charge, true even before Copernicus met opposition in seeking to remove the earth from the center of the universe. However, discoveries of the past two decades have removed much opposition and turned most of the old skeptics into believers.

The subject of continental drift and polar wandering has effected a revolution in the earth sciences during the past decade. That revolution is still under way. No single question has ever before prompted scholars from so many specialties in earth science to unite in an attempt to provide answers. The past decade has produced very strong evidence in support of continental drift, largely from studies of the ocean floor and ancient rock magnetism, but the question of former continental connections was raised a long time ago. Let us begin at the beginning.

Although Sir Francis Bacon is often mentioned as the first to comment on the subject of drift in 1620, based on the similarity of the opposing shores of the Atlantic continents, he really spoke of the similarity of the corresponding west coasts and failed to compare the coasts facing each other across the Atlantic. Furthermore, he never clearly indicated that the old and new worlds had once been connected.

Father Francis Placet touched on the subject of continental separation but did not discuss drift. In a classical, heavy-handed, religious work of 1668, he used the then widely accepted ideas of the biblical flood and vertical earth movements to split continents. Like Bacon before him, he too did not really discuss the subject of "drift." Alexander von Humboldt, unlike Placet and Bacon, did comment on the similarities between the opposite coasts of the Atlantic in 1801, but he explained their separation by erosion.

It was not until 1858 that the subject of continental drift was opened in the true sense by Antonio Snider-Pellegrini in his **La Création et ses mystères**

dévoilés. He illustrated his concept of separation and sidewise movement with two maps (fig. 9.1) showing the reassembly of the continents for late Carboniferous time (Pennsylvanian period, about 325–270 million years ago). Snider-Pellegrini's argument for former connection of the continents was based on the very great similarity between large numbers of fossil plants in the Carboniferous-age coal measures of Europe and those of equivalent age in North America. Snider-Pellegrini appears to be the first author of significance on the subject of drift and probably provided the germ seed which was later cultivated by Taylor and Wegener in their almost simultaneous, but independent, presentations in the first decade of the twentieth century.

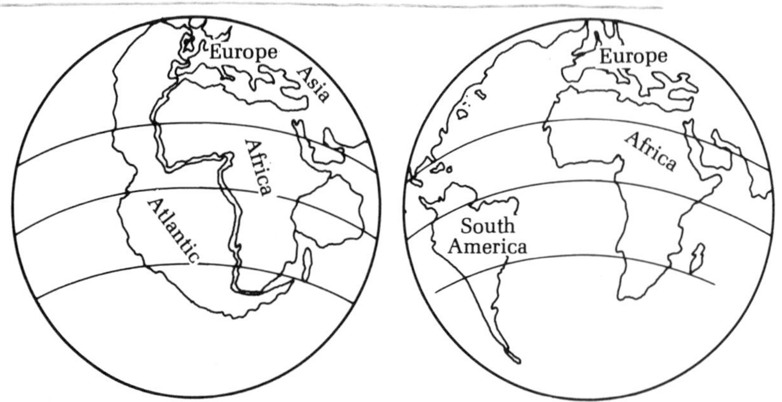

FIGURE 9.1 Antonio Snider-Pellegrini's 1858 Reassembly of the Continents for the Pennsylvanian Period, about 300 Million Years Ago, Appears on the Left. It is based mainly on the similarity of fossil plants of Pennsylvanian age found in common between Europe and North America.

Edward Suess of Vienna in 1885 published **The Face of the Earth,** a lengthy work of several volumes in which he coined the term **Gondwanaland** for a hypothetical supercontinent that included India, Australia, Antarctica, and parts of southern Africa and South America all joined together by land bridges. Gondwanaland was named after Gondwana, a province in east central India where the rocks include unusual fossil plants and evidence of extensive glaciation (late Paleozoic and early Mesozoic age, about 325–225 million years old). Each of the continents which Suess joined in his jigsaw puzzle has both the unusual glacial rocks and the fossil plants found in the Gondwana region. Suess did not resort to drifting of the continents but instead advanced the then widely held theory of land bridges to account for the intercontinental similarities in rocks and fossils. The land-bridge proponents believed that extensive landmasses connecting the present continents have sunk beneath the ocean.

Although the earliest comprehensive treatment of the subject, including suggestions for ways in which the continents could have been moved great distances, was done by Alfred Wegener in 1910, he does not hold priority. Two years earlier, in 1908, F. B. Taylor, an American, privately printed a pamphlet in which he advanced ideas to account for the present distribution of mountain ranges. He saw the continents as "huge landslides from the polar regions towards the equator," both in the Northern and Southern hemispheres, and envisioned the original great continents as continuous sheets of rock which ripped apart with the pieces moving from the polar regions toward the equator.

Taylor believed in two major ideas which have been strongly opposed: first, the mountain structures which he said were produced by thousands of miles of sidewise movement could have been produced by only a few tens of miles of displacement — it seemed unnecessary to move the continents thousands of miles to build mountains; second, he proposed that the moon became a companion of the earth about 100 million years ago (during the Cretaceous period). During that early time, gigantic tidal forces slowed the rate of earth rotation and pulled the continents away from the poles. Geologists objected, since the Cretaceous capture of the moon by the earth left no explanation for those mountain ranges formed before Cretaceous time. Jeffreys also objected on the grounds that if the new tidal force were strong enough to raise mountains, then it would have stopped the earth's rotation.

In 1911, H. H. Baker, an American, proposed that a great supercontinent was split along the Atlantic and Arctic oceans at the close of the Miocene epoch (about 7 or 8 million years ago). He rotated certain areas in order to improve the jigsaw fit of the continental margins. Baker also suggested that the orbits of the earth and Venus had varied, bringing them closer together to produce great tidal forces which tore a piece of earth from the Pacific Ocean area to form the moon. The earth's crust then supposedly broke up, and pieces slid toward the Pacific basin to fill the void. Modern geologic evidence for such a catastrophic event is lacking.

Lasting worldwide attention was focused on the question of drifting continents in 1912 by Alfred Wegener. His monumental book, **The Origin of Continents and Oceans,** was first published in 1915 with the fourth and last edition appearing in 1928. It was more comprehensive and detailed than Taylor's and drew from almost all lines of evidence which were available at the time. Wegener's ingenuity and perspective for synthesis are best judged against the geological backdrop of the time in which he wrote and the storm of controversy which he precipitated and became the center of. He unfortunately died a tragic death of starvation on the Greenland ice cap while on an expedition in 1930. Wegener portrayed his grand concept in maps (see figs. 3.1 and 9.2) which have since become familiar, classic reproductions. Wegener proposed that all of the continents were united in a single landmass called **Pangaea** (**pan**, all; **gaea,** earth, in Greek), which was torn apart during the Mesozoic era (about 225–70 million years ago). The southern continents were pulled away first; they were followed by the westward drift of North America, and ultimately Greenland departed. Like Taylor, Wegener saw the continents undergoing **Polflucht** — the flight from the poles. He described the westward drift of the Americas as producing the crumpling of the western mountain ranges as they met resistance from the Pacific Ocean floor.

The westward-floating Americas simultaneously left a trail of debris in the form of the West Indies. Similar descriptions accompanied his explanations of movements of the other continents; he attributed movement to the attraction of

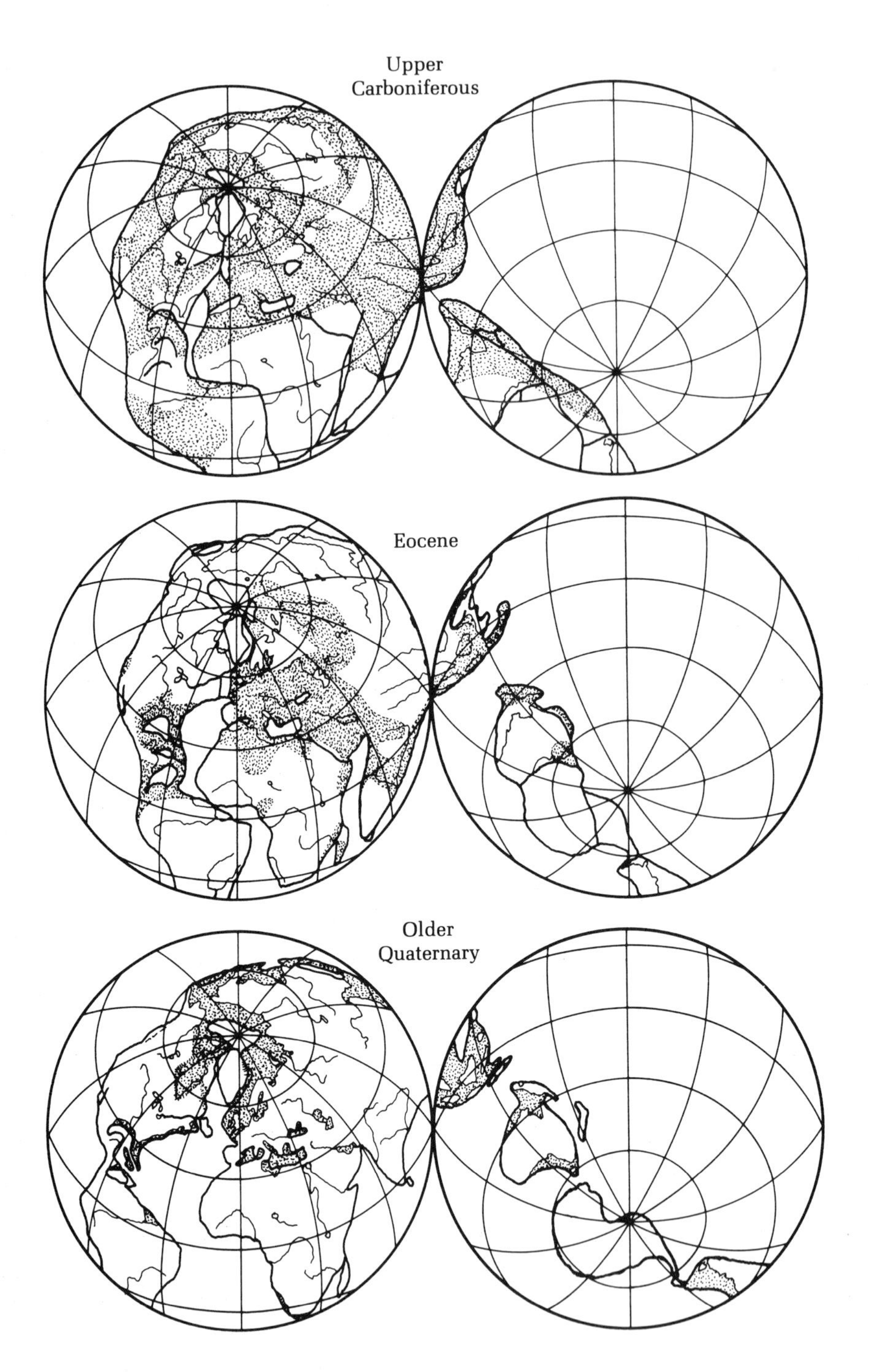

FIGURE 9.2 Wegener's Reconstructions of the Map of the World for Three Periods according to his Continental Displacement Theory. The same reconstructions are shown in another map view in fig. 3.1. (Reprinted, by permission, from Alfred Wegener, *The Origin of Continents and Oceans*. English translation of the 3rd German edition, 1924. London: Associated Book Publishers, Ltd., 1924.)

the moon and sun on the continents. Wegener also noted the following in the opening of his book:

> He who examines the opposite coasts of the South Atlantic Ocean must be somewhat struck by the similarity of the shapes of the coast lines of Brazil and Africa. Not only does the great right-angled bend formed by the Brazilian coast at Cape San Roque find its exact counterpart in the reentrant angle of the African coastline near the Cameroons, but also, south of these two corresponding points, every projection on the Brazilian side corresponds to a similar shaped bay in the African. Experiment with a compass on a globe shows that their dimensions agree accurately.
>
> This phenomenon was the starting point of a new conception of the nature of the earth's crust and of the movements occurring therein; this new idea is called the theory of the displacement of continents, or more shortly, the displacement theory, since its most prominent component is the assumption of great horizontal drifting movements which the continental blocks underwent in the course of geological time and which presumably continue even today.

Wegener moved from the congruence of the South Atlantic coastlines to refuting the then widely prevalent contraction theory which held that the earth has shrunk by cooling, which produced the major folded mountain ranges. He also denied the idea that land bridges across the ocean deeps had sunk to form the present ocean floor. Intercontinental matching of rock types, major fault zones, and mountain structures were treated in depth. He touched on all lines of appropriate evidence which were available in 1912 and laid the cornerstone for one of the major theories in the natural sciences. This broadly inclusive theory has revitalized many old disciplines and led to the integration of other formerly isolated subject areas in order to focus on so grand a theme. Wegener may rightfully be considered the father of continental drift theory.

Following publication of Wegener's theory, many scientists fell to serious consideration of the question, and some came to believe that there were two original supercontinents, rather than one as Wegener had postulated; one lay in the Northern Hemisphere and the other in the Southern. Of this group, Alexander Du Toit, a Johannesburg, South Africa, University professor, proposed the most significant alterations to Wegener's ideas. In **Our Wandering Continents,** of 1937, he concentrated on Permian-age (270–225 million years ago) glacial deposits found in the southern continents and contrasted these with coal deposits of the same age (indicating tropical conditions for that time) found as far north as Spitzbergen in the continents of the Northern Hemisphere.

To "resolve this climatic paradox" (of glaciers in the southern continents and tropical coal-forming swamps in the northern continents at the same time during the Permian), Du Toit brought the southern continents together at the South Pole and arranged the northern continents in another group, such that the coal forest deposits were located at the equator. He retained the name **Gondwanaland,** coined earlier by Suess, for the southern supercontinent but invented the term **Laurasia** for the northern supercontinent (see fig. 9.3).

Du Toit offered a new force by which continents moved (see fig. 9.4): (a) sediments (sand, silt, clay, etc.) that have been eroded from the continents are laid down in layers of great thickness, in the shallow seas bordering the continents; (b) as the sediment piles up, the floor of the sea along the edge of the continent sinks such that very great thicknesses of beds develop over tens of

FIGURE 9.3 Du Toit's Setting of the Continents at the Close of the Carboniferous (about 280 Million Yr Ago). The northern continents are grouped as Laurasia, and the southern ones form Gondwanaland. He noted that no allowance had been made for Paleozoic mountain-building compressions of the crust which would alter the position of Laurasia. (Reprinted, by permission, from A. L. Du Toit, *Our Wandering Continents*. Westport, Connecticut: Greenwood Press, 1937.)

millions of years; (c) this sinking of the sea floor along the edge of the continent tilts the continent, resulting in its slide toward the sea; (d) the movement of the coastal part of the continent toward the sea results in the stretching and eventual tearing of the middle part of the continent; (e) molten rock (magma) then forces its way, under great pressure, upward into the torn zone and adds to the seaward movement of part of the continent. By this device involving erosion, deposition, sinking of parts of the shallow sea floor, earth gravity, and molten rock forces, the continental crust is forced to move over the underlying mantle (the denser layer beneath the crust, composed of rock richer in iron and magnesium than the crust itself).

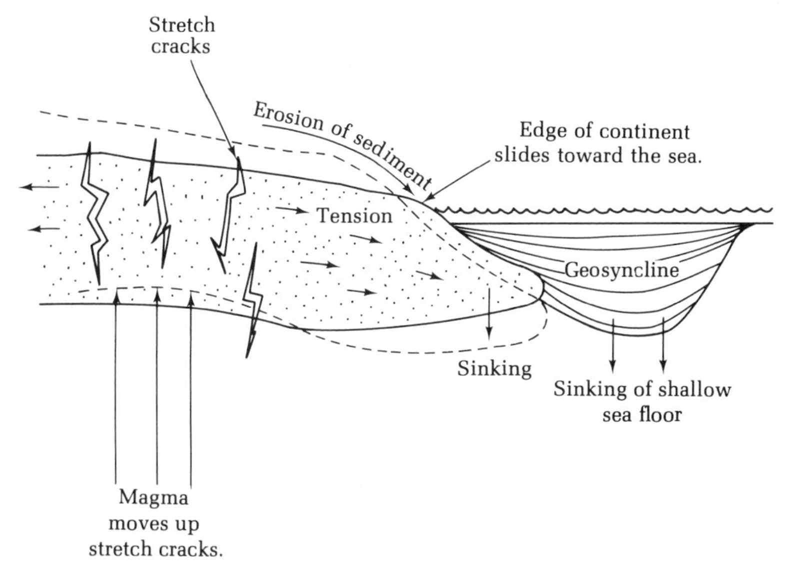

FIGURE 9.4 Du Toit's 1937 Idea of How Continents Move. (Reprinted, by permission, from A. L. Du Toit, *Our Wandering Continents.* Westport, Connecticut: Greenwood Press, 1937.)

During this period of the 1920s and 1930s, strong opposition to drift was voiced from many quarters; especially notable was a symposium on the subject held in 1928 in New York sponsored by the American Association of Petroleum Geologists. Most of the symposium contributors were strongly opposed to the concept of drift on many grounds, including some of the following:

(1) Sir Harold Jeffreys of England believed that the crust and mantle are too rigid to permit large-scale movement of the continents. He argued that no force is known which is great enough to move the continents.
(2) While Wegener believed the original supercontinent did not split until the Mesozoic era (about 225 million years ago), the opponents saw no reason why drifting should not have occurred much earlier in earth history.
(3) The jigsaw-puzzle fit of the continents was said to be far from perfect.
(4) Wegener's evidence of great similarity in animal species between the old world and new was said to be greatly overstated.

Although the symposium was intended to be a balanced pro-and-con treatment of the subject, and many new, important ideas were contributed, the net effect was surely negative and further disposed most of the professional community toward rejection of the drift concept.

Convection was first recognized by Count Rumford in 1797. The possibility of convection below the crust was first suggested by William Hopkins in 1839, but it was not until 1881 that convection was applied by Osmond Fisher in a specific way to explain mountain building. He believed that the drag of such currents on the underside of the crust near the continental margins caused local concentrations of heat such as to produce volcanoes.

The Austrian geologist Otto Ampferer also suggested, in 1906, that mountains are folded and raised by the drag of convection currents in the viscous material of the mantle; Ampferer erroneously believed that the mantle was in a liquid state — unaware that observations of earthquake waves in 1899 had indicated that the mantle is solid.

John Joly, an Irish physicist, revived and elaborated Ampferer's convection theory in 1926 to explain the time schedule of mountain-building movements of the earth's crust; like Ampferer, Joly too thought the mantle was liquid. He believed that (1) heat in the earth is produced by radioactivity; (2) because the crust is a poor conductor, it traps the heat produced inside the earth; (3) the heat accumulates just below the crust and melts the mantle rock; (4) the mantle expands and becomes less dense; (5) the crust which was floating on the mantle sinks into it as deep as one mile; (6) when the mantle is partly melted, a portion of it undergoes convection. Joly employed the theory of convection circulation in the mantle in order to explain periodic mountain building; he was not originally concerned with continental drift.

The "drifters" welcomed Joly's convection hypothesis and were quick to modify and utilize it. Only two years after Joly's proposal, Arthur Holmes of Edinburgh University first suggested that the solid, as well as fluid, material of the mantle could undergo convection circulation. Holmes' concept of convection circulation within the solid mantle involved the following steps: (a) a current rises vertically below the central part of a continent; (b) the current forks to produce two horizontal currents moving in opposite directions at the top of the mantle; (c) the opposed horizontal currents split the continent and carry off their respective parts; (d) each of the horizontal currents collides with currents from other heat sources and is deflected downward, pulling the continental margin into a downward bulge; (e) when the convection current subsides, the continental crust which had been bulged downward will be raised by simple flotational force, much as a cork will surface after being held below the surface of water.

Holmes' convection theory of 1928 was similar to the presently accepted one (see fig. 5.10). In summarizing his discussion, he noted that "during large-scale convection circulation the basaltic layer [dark, dense volcanic rock of the ocean floor] becomes a kind of endless traveling belt on top of which a continent can be carried along, until it comes to rest [relative to the belt] when its advancing front reaches the place where the belt turns downwards and disappears into the earth." Although this theory was first proposed in the "Transactions of the Geological Society of Glasgow" in 1928, it went virtually unnoticed until this decade. In the same year that Holmes made his presentation on convection, the eminent R. T. Chamberlain of the University of Chicago condemned the continental drift concept on eighteen different counts and was thereafter heard to speak of it only in "scoffing terms" according to Chester Longwell of Yale University.

David T. Griggs of U. C. L. A. agreed with Holmes' concept and made model experiments of convection in 1939 which included four stages and were similar to Joly's in many ways. The model experiments of Griggs probably were prompted in part by and were also in agreement with the results of the Dutch geophysicist, Felix A. Vening-Meinesz, who, in the early 1930s, ingeniously took an instrument to measure gravity (which varies slightly from place to place on the earth) to sea in a submarine. The submerged submarine provided a steady platform from which to measure gravity. The results indicated that the island arcs of the East Indies and the western Pacific were areas of deficient or very low gravity. Vening-Meinesz concluded that therefore the earth's crust was not in a state of flotational balance (isostasy), but instead that some force operating there drags the crust downward to form deep-sea trenches and holds it there against the flotational force of gravity. Such a force was holding the earth's surface in its irregular shape, maintaining deep-ocean trenches and high mountains against the natural tendency of gravity to flatten it out. The convection current hypothesis of Holmes and Griggs was certainly in agreement with the unusual gravity observations of Vening-Meinesz.

During the mid-1930s, additional theoretical work on the rate and stresses involved in convection was done by C. L. Pekeris and A. Hales. The notion of convection currents inside the earth became generally accepted by the early 1950s. In 1949 John Verhoogen of the University of California at Berkeley noted that a rotating body such as the earth with internal sources of heat theoretically could not be in a state of liquid balance with no internal flow; he suggested in 1954 that convection would explain adequately the generation of molten rock within the earth. Bullard and others of the fifties sought to show that convection in the earth's core also indicated such movement in the mantle in order to carry heat away from the core. The limits of convection were still later defined by Runcorn, Chaudrasekhar, and others. In spite of the accumulating evidence, the renowned Baily Willis of Stanford, certainly speaking for a majority of the professional community, suggested as late as 1944 that the hypothesis of continental drift in effect be outlawed as an obstruction to progress in an article titled "Continental Drift, Ein Märchen [a fairy tale]" in the **American Journal of Science**.

An alternative to convection as the engine of drift was advanced in 1963 by J. Tuzo Wilson who proposed that there are hot spots or rising plumes which originate deep in the mantle and remain fixed for 100 or more million years. In 1972 W. J. Morgan expanded this idea, theorizing that the hot spots raise the lithosphere above them into domes 2 or 3 miles high and that a line of such domes may become connected by a mid-oceanic ridge rift zone. Then the crustal plates merely slide off the dome regions in either direction under the force of gravity to drift over the weak, semimolten asthenosphere below them.

A new area of investigation which only indirectly lent itself to support of the drift hypothesis came from those who were actively mapping the ocean floors during the 1950s and 1960s. A Columbia University group which included Maurice Ewing and Bruce C. Heezen, and H. W. Menard of the University of California, were active in demonstrating that only a very thin layer of sediment covers the ocean floors. They reasoned that if oceans have been receiving sediment washed from the continents throughout geological time, they should contain sedimentary layers very much thicker than those actually present.

If sediment had been accumulating throughout geologic time, to where did it all disappear during the last 100 or so million years? The sediment which they found accounted for deposition only since the Cretaceous period (about 100 million years ago). They further discovered that all of the rocks dredged from the ocean floor were geologically quite young (less than 200 million years). This is difficult to explain since it is widely held that the earth's continental and oceanic history covers about 3000 million years! The ocean floor researchers also found that all islands and submerged volcanoes of the ocean were geologically very young. Why were no volcanoes or islands present from early in earth history?

Perhaps most significant of all, the deep-sea researchers of the 1950s and 1960s clearly laid out the major systems of submarine mountains and faults which led to other crucial lines of research. These oceanographers had mapped the longest continuous range of mountains on the earth rising from the ocean floor. The range is about 40,000 miles long and 300 to 3000 miles wide. In places, they rise to 10,000 feet above the ocean floor, forming islands at their high points. In some places the mid-oceanic ridges go inland and become great fault zones on the continents. Also discovered were great fracture zones and faults which are associated with the mid-oceanic ridges (see fig. 7.4).

These ocean floor discoveries of the 1950s sparked new interest and led to what might be called part two of our brief history of the subject – the advent of new pieces of evidence which have virtually confirmed the idea that the continents have drifted and the poles have changed their positions.

Several modified versions of Holmes' convection hypothesis were offered that incorporated more recent evidence; the most notable was that of Harry H. Hess of Princeton University, who in 1962 eloquently expanded much of Holmes' original hypothesis in the course of a paper titled "History of the Ocean Basins." Hess considered his paper an "essay in geopoetry." Evidence accumulated since then has substantiated his profound poetical speculations. He suggested that the ocean floor is moving because convection currents rise directly beneath the mid-oceanic ridges, split, and spread apart, carrying the ocean floor away from the ridge in opposite directions. New ocean floor is created at the ridges by rising lava. The floor is moved horizontally away from the ridge such that the leading edges are dragged down into the deep-sea trenches at the edge of the continents. Of course, the older part of the ocean floor constantly is being destroyed and would account for the absence of the older sediments and islands. This idea fits well in the Pacific, but the Atlantic lacks the deep-sea trenches. Robert S. Dietz of the U. S. Coast and Geodetic Survey also reconsidered and expanded the Holmes hypothesis in 1961 and applied the term **sea-floor spreading.**

A significant difference between Holmes' theory and recent versions is that Holmes showed fragments of the continents which had been left behind "floating" directly above the rising parts of the convection current, where it splits and diverges under the center of the mid-oceanic ridge; modern versions have no such lagging pieces of continental crust in the area of supposed stagnation, just above where the rising convection column diverges.

Heat flows upward from the earth's interior; it has been measured in oil-well drilling, mines, and tunnels and generally has been found to increase with depth. The first measurement of heat flow from the ocean floors was done by Sir Edward Bullard and A. E. Maxwell and their colleagues with an instrument which they devised at the Scripps Institution of Oceanography. In 1954, while examining the results obtained on board the research vessel **Discovery II,** Bullard found that the heat flowing from the ocean floor along the crest of the Mid-Atlantic Ridge was two to eight times the heat flow found elsewhere on the ocean floor or the continents. Such unusually high heat flow further suggested that the mid-oceanic ridge had formed directly over the hot, rising part of a convection current. Further heat flow measurements by R. P. Von Herzen, in 1956, over the East Pacific Ridge showed similar, unusually high heat-flow values. The heat flow decreased with increasing distance away from the ridge and gave very low readings near the middle America trench. Low heat flow should be expected over the deep-sea trenches, since they supposedly are located over the point at which the convection current turns downward and drags the crust with it, forming such trenches in the crust.

This idea of downward drag is further supported by data from earthquake-recording instruments (seismographs); when the points of origin (foci) of deep earthquakes are mapped, they usually are found to lie in a plane which extends downward at about 45° from the floor of a deep-sea trench. It is thought that the sharp movement of earth material along this plane causes earthquakes. From 1957 to 1967 almost all earthquakes plotted by the U. S. Coast and Geodetic Survey were located along the deep-sea trenches or the mid-oceanic ridges, the most likely places for quakes if the sea floor is in motion.

About 1960, in the midst of all this new speculation on drift and the Holmes hypothesis, Ronald G. Mason, Arthur D. Roff, and Victor Vacquier of the Scripps Institution of Oceanography announced that the floor of the ocean off the North American west coast showed an extraordinary pattern of variations in the magnetic intensities of the rocks composing it. The pattern showed up as a series of parallel rock stripes. The series of magnetic stripes ran roughly north-south and were broken and offset along lines at a right angle to the stripes (roughly east-west). The stripes went unexplained until 1963 when F. J. Vine (now at Princeton) and D. H. Mathews of Cambridge University suggested that if lava were extruded along the center of the mid-oceanic ridges, the iron in it would become magnetized in the direction of the earth's magnetic field which was present at the time of cooling and would then be carried away from the ridge center. This theory was also proposed independently in 1964 by Morley and Larochelle.

Knowledge that certain rocks and baked pottery could be magnetized by the earth's magnetic field and retain their magnetism had been discovered in the last century, and research was later conducted in several countries during this century; significant work on paleomagnetism was done by Nagata and students in Japan and Thellier in France in the 1930s, but it was not until the late 1940s that L. Neel, a French physicist, explained how igneous rocks can become magnetized in the earth's magnetic field and explained that such magnetism does not change when the magnetic field of the earth changes. The first rocks showing **reversed** magnetism were discovered in France in 1906 by Brunhes and David, giving rise to the question of whether the earth's magnetic field had been reversed in the past in order to produce such reversely magnetized rocks. Mercanton in 1926 first argued that the earth's magnetic field had reversed itself, and he encouraged workers on other continents to obtain rock samples showing such reversals. He clearly stated that data on ancient rock magnetism might allow us to test the hypothesis of polar wandering and continental drift. M. Matuyama of Kyoto Imperial University, after examining magnetized rock samples from many areas in Japan, noted the connection between rock age and the direction of its magnetization. He estimated the age of a magnetic reversal for the first time in 1929. Almost all magnetic evidence now shows that the earth's magnetic field has indeed reversed itself repeatedly throughout geologic time. North has become south; south has become north.

Vine and Mathews were searching for evidence of reversals of the earth's ancient magnetic field. Rocks on land had already shown such magnetic reversals. The land rocks indicated that the earth's magnetic field unexplainedly had reversed itself repeatedly in the past.

They believed that the lava being erupted along the mid-oceanic ridge crest would become magnetized in the direction of the prevailing earth magnetic field; as the sea floor spread, it would carry that rock away from the ridge center in both directions. As new material erupted at a later date along the ridge center, it would in turn be magnetized with a possibly reversed magnetism. If this process had actually taken place repeatedly, they hoped to find alternating bands or strips of rock showing normal and reversed magnetism.

The exact dating and counting of magnetic reversals recorded in rocks on land does not extend back more than about 6 million years. Magnetic reversals in older rocks have been encountered, but they do not lend themselves to dating or counting. All older reversals therefore are based on evidence from the ocean floors.

While Vine and Mathews were suggesting a search for such alternately magnetized rock bands, J. R. Heirtzler and a research group from the Lamont-Doherty Geological Observatory of Columbia University and the U. S. Naval Oceanographic office were conducting just such a magnetic survey of the Reykjanes Ridge in the North Atlantic, a part of the mid-oceanic ridge. Towing sensitive magnetometers behind an ocean research vessel, they surveyed the floor of the ocean and found that magnetic stripes were distributed parallel to the axis or center of the ridge in a symmetrical way such that the series of stripes on one side of the ridge were a mirror image of the pattern on the other side of the ridge (see fig. 7.6b).

Vine and Mathews found what may be regarded as the important key to sea-floor spreading. In 1963 they showed that ratios of the width of the magnetically striped rocks of opposed polarity on the sea floor are the same as the ratios of polarity intervals observed in young rocks on land. This ratio is the key to interpreting spreading rates. The work of A. Cox and R. R. Doell and others during the 1950s on the dating of polar reversals recorded in young rocks on the continents really provided the necessary base on which Vine and Mathews worked. Shortly afterward, F. J. Vine and J. Tuzo Wilson demonstrated that the striped pattern of rocks showing magnetic reversal from the Reykjanes Ridge south of Iceland matched the pattern of magnetic reversals found on the sea floor off the North American west coast.

In 1966 Opdyke's group discovered that the weak magnetization found in deep-sea cores showed a succession of polarity reversals with depth which was identical to those found on land and which also agreed with the ratios of the width of magnetically striped rocks of the sea floor. Therefore, three different sets of magnetic data derived from three different features were shown to have identical ratios; the generalizations derived have facilitated a number of accurate predictions about spreading rates which are in agreement with estimates from other sources. In 1968 J. R. Heirtzler and his associates constructed a time scale for magnetization of ocean-floor rocks extending back through Cenozoic time. It was based on surveys of magnetic anomalies over mid-oceanic ridges in different oceans; the same pattern of anomalies appeared in all of them. The Heirtzler time scale was supported shortly afterward by data gathered by the J. O. I. D. E. S. Sediment cores from the South Atlantic dated by microfossils which showed increasing age with increasing distance from the ridge crest. The cores indicated a sea-floor spreading rate of about 2 cm/yr (4/5 in) which was in close agreement with the spreading rate implied by the Heirtzler time scale.

The continually spreading sea floor may be thought of as a conveyor belt acting as a magnetic tape recorder which allows us to play back the history of the earth's changing magnetic field.

Let us now return to the early 1950s and look in on groups of English, Japanese, French, and U. S. scientists who were at work refining the earlier ideas on ancient or fossil rock magnetism which had been advanced by L. Neel in the late 1940s. Some of Neel's ideas were prompted by questions raised by John Graham of the Department of Terrestrial Magnetism in the United States.

After more than twenty years of virtual stagnation, the continental drift question was resurrected by the ancient-rock magnetists. It had been shown earlier that if eruptions of lava took place today on three different continents, the rocks formed would be like three different compasses with their needles frozen in place pointing toward the present magnetic pole. Using such knowledge, in 1953 a group at London University, including the Nobel laureate P. M. S. Blackett, found that English, Triassic-age (about 225–180 million years old) rocks were not magnetically lined up with the earth's present magnetic field. The paleomagnetists found it necessary to rotate England 30° clockwise and move it to a lower latitude to make order of the magnetic directions found in the Triassic-age rocks. The London group then measured the ancient magnetism of the Deccan Plateau basalt rocks of India which had formed during the last 180 million years and found that it was necessary to move India to the Southern Hemisphere in order to satisfy their ancient rock magnetism data.

At about this same time, another English group, under S. K. Runcorn who was then at Cambridge University, was measuring ancient magnetism in rocks all over Europe. Runcorn and workers from Newcastle University drew paths for the poles as indicated by evidence from the different continents and found that they did not coincide. The "polar-wandering" paths of Runcorn and others do not necessarily mean that the poles have moved; such "polar-wandering" curves may simply indicate that the continents have been moving relative to a fixed pole. Inasmuch as the polar curves are different for each continent, then the continents must have moved relative to each other. It is generally believed that the earth's magnetic and rotational poles have always been very close to each other — within about 15°.

This new question of paleomagnetism raised by the English in connection with the problem of continental drift not only helped to drag continental drift out of the closet after more than twenty years, but also gave birth to the question of magnetic pole movement. By the middle 1950s, Wegener's hypothesis was ready to race again. It was the magnetists who brought forth information which could be expressed in numbers in a concrete way. The snickering of the antidrifters began to subside.

Crustal Plate Theory

By the mid-1960s, the explosion of knowledge about the question of continental drift had produced some unique, new ideas and many revised versions of old ones. This new knowledge pointed to the interconnectedness of such features as major faults on the land and in the oceans, continental drift, and spreading sea floors.

In 1965, J. Tuzo Wilson of the University of Toronto sought to explain the abrupt ending of such major features of the crust as mountains, mid-oceanic ridges and major faults with great horizontal displacement; most such features end beneath the oceans. Wilson suggested that they are really not isolated with abrupt ends, but instead are connected, forming a network of mobile belts within the crust. He believed that these belts mark the boundaries of several great rigid crustal **plates** which fit together to form the entire earth surface. Thus was presented the first notion of plate tectonics, a contribution which is gathering increasing historic importance (see fig. 9.5).

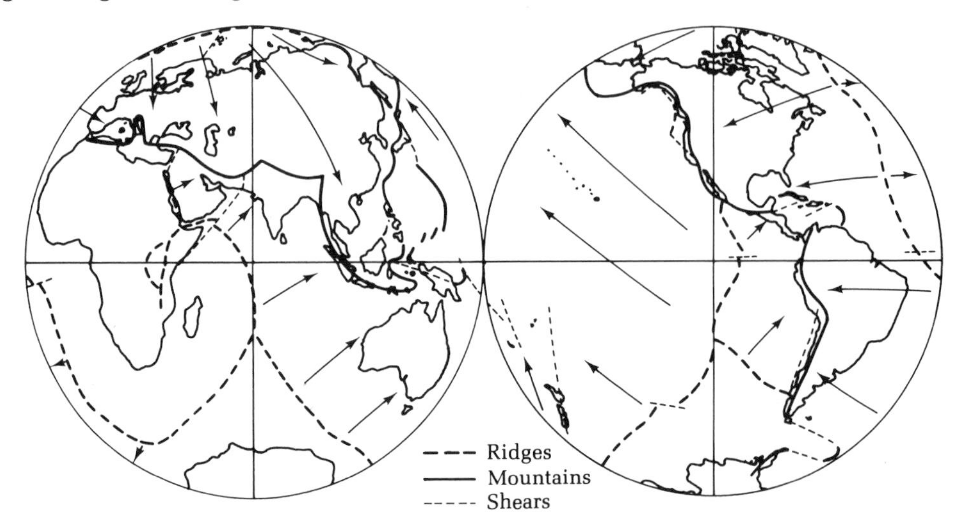

FIGURE 9.5 Wilson's Monumental Sketch Map Showing the Network of Mobile Belts around the Earth Which Are the Boundaries of the Several Large, Rigid Plates Forming the Earth's Surface. Mountains and island arcs are formed by compression (solid lines), active transform faults along which plates slide past each other (light dashed lines), and active mid-oceanic ridges, which are caused by tension (heavy dashed lines). [Reprinted, by permission, from J. T. Wilson, Nature 207 (1965): 343–47.]

Wilson's main idea was that any feature such as a fault or a mid-oceanic ridge or a mountain range could be **transformed** into a different feature at its apparent end point. He maintained that one type of motion which deforms the crust could be changed or transformed into a different type of motion across the junction of two different structures. He called the junction at which one feature (such as a fault) changes into another (such as a mid-oceanic ridge) a **transform.** He noted in closing that proof of the existence of transform faults would aid greatly in "establishing the reality of continental drift" and in showing how such movement has taken place.

In 1966, a year after Wilson's contribution of the transform fault concept, L. R. Sykes, a seismologist at the Lamont-Doherty Observatory, analyzed shock waves produced by earthquakes along faults which offset the mid-oceanic ridges. His study indicated that the motion on such faults was as had been suggested by Wilson. Thus the transform fault concept moved from speculation to fact. In the same year, Fred Vine importantly summarized the available data on the rates of ocean-floor spreading and showed that the rates (for one side of a ridge) varied greatly from almost 2 inches (4.5 cm) per year in the south Pacific to only 2/5 inch (1 cm) per year on the Reykjanes Ridge, southwest of Iceland. The contributions of 1965 and 1966 engendered a major change in attitude toward the drift-plate tectonics question; many earth scientists turned to examine it with a new eye — but perhaps not all. In 1966 **The Source Book of Geology: 1900–1950** was published; it is a compilation of the important concepts contributed during the first half-century. No mention of the hypothesis of continental drift nor its author, Alfred Wegener, is made in the entire volume.

In March, 1968, W. J. Morgan of Princeton University extended the transform fault concept of Wilson and divided the earth's surface into about twenty rigid blocks of crust. He assumed that each block is "bounded by rises [where new surface is formed], trenches or young fold mountains [where surface is being

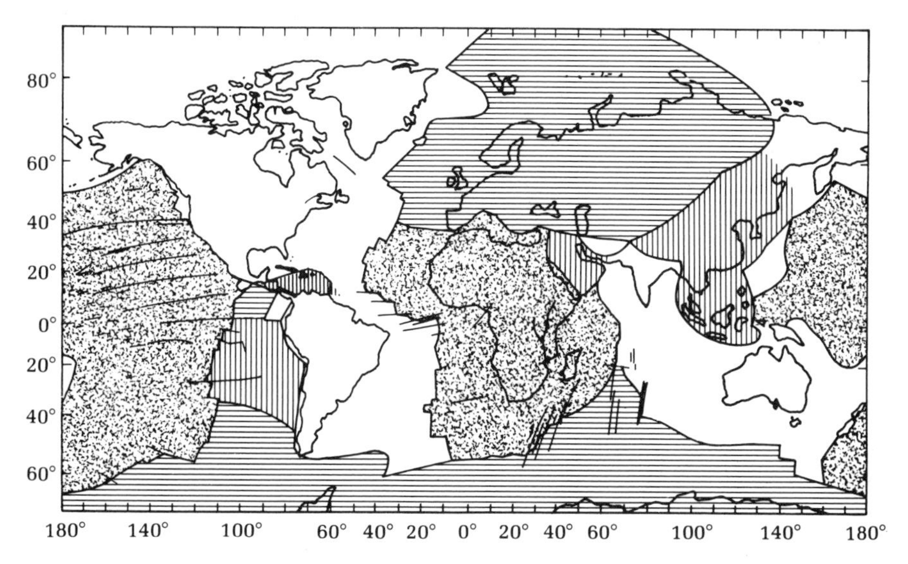

FIGURE 9.6 W. J. Morgan's Map of the Earth's Crust Divided into Moving Rigid Slabs. The boundaries between the blocks (plates) are rises (mid-oceanic ridges), trenches or young folded mountains, and transform faults. His figure was based on Sykes' 1968 ridge-system map and Heezen and Tharp's tectonic map of 1965. (W. J. Morgan, *Journal of Geophysical Research* 73, p. 1960, 1968, copyright by American Geophysical Union.)

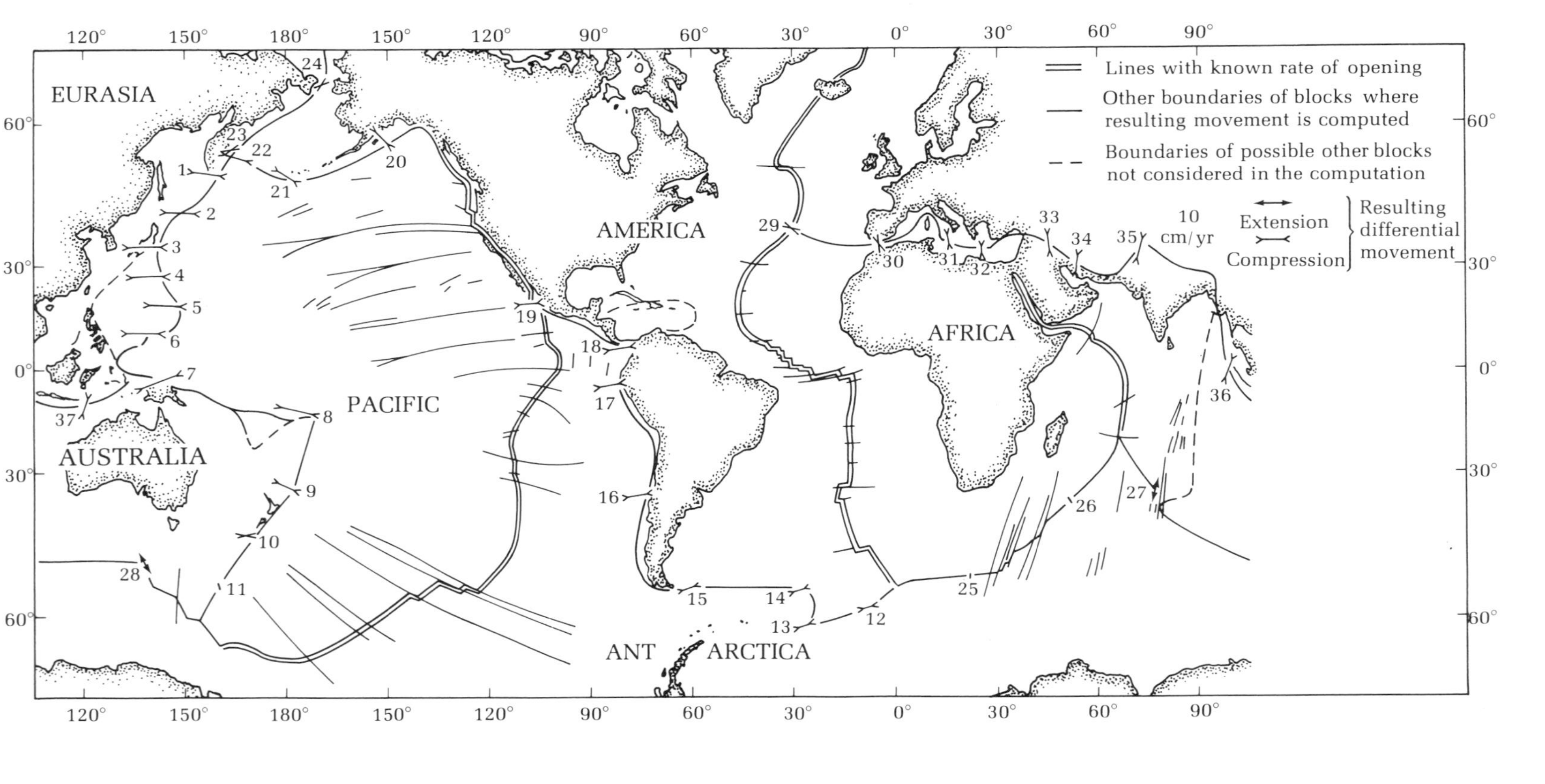

FIGURE 9.7 X. LePichon's World Map Showing the Locations of the Boundaries between the Six Blocks Which He Used in His Computations Which Emphasized Differential Movements in Belts of Crustal Compression. (X. Le Pichon, *Journal of Geophysical Research* 73, p. 3675, 1968, copyright by American Geophysical Union.)

destroyed], and great faults, and that there is no stretching, folding or distortion of any kind within a given block" (see fig. 9.6).

In June of 1968, X. Le Pichon of Columbia University supported Wilson and Morgan's view of an earth crust divided into blocks or plates. His support was a geometric analysis of the data which had been accumulated from many sources bearing on sea-floor spreading. Le Pichon's hypothesis included six rigid crustal plates which cover the earth's surface, the area of which remains unchanged (see fig. 9.7). The plates or blocks move about like rafts. An ocean ridge forms where two blocks move away from each other; new crust is formed in the zone of departure. When two blocks move toward each other and collide, compression produces a folded mountain region, or part of one block is thrust beneath the other and crust is destroyed.

B. Isacks, J. Oliver, and L. R. Sykes of Columbia University, in September of 1968, coined the term **new global tectonics** to cover the new broad view of deformation of the earth's crust which included continental drift, sea-floor spreading, and transform faults and their relationships to earthquake shock wave studies. Their studies of earthquake waves led them to support Le Pichon's belief in a crust composed of six rigid blocks or plates. This new broad view of worldwide crustal features and the forces which have deformed them is now generally referred to as **plate tectonics.**

The plates are about 60 miles thick, composed of the crust and the upper part of the mantle. The American plate includes North and South America and the western part of the Atlantic Ocean floor. The African plate includes the eastern half of the South Atlantic floor, part of the Indian Ocean and the whole African continent. The Eurasian block does not include India and Arabia. Australia and Antarctica rest on separate blocks and the Pacific Ocean is carried on a plate which includes no land. When two plates carrying continents crash, the collision folds the land into mountains such as the Himalayas, which resulted from the collision of nothward-moving India with Asia. Because continental crust is lighter than oceanic crust, when the two collide, the oceanic plate is thrust under the more buoyant continental one. Such a meeting results in ocean trenches and mountains such as found along South America's west coast. Here the Peru-Chile trench and the Andes Mountains have formed where the Pacific plate dives under the American plate. When two plates strike each other at an angle rather than head on, a major fault results, such as the San Andreas in California, where the Pacific plate slides past the American plate along the fault, producing periodic earthquakes. In general, earthquakes are distributed almost exclusively in the boundary areas between plates. Generally, shallow quakes occur along the ocean ridges. Deep-seated and shallow quakes occur in the arcs, deep-sea trenches, and great youthful mountains such as the Himalayas.

Plate tectonics calls upon convection to bring up new hot material from the mantle to form new crust along the mid-oceanic ridges. Old crust dives to disappear somewhere in the mantle at the places of plate collision. The remaining great question is, Do convection currents cause the motion of the plates, or does the plate motion drive the convection currents? The basic force which propels the plates is still not understood. Do eruptions take place along the mid-oceanic ridges because the plates separate to provide an opening, or does the rising, erupting material force the plates apart and drive them?

Several workers have done elaborate reconstructions of the orginial protocontinent or protcontinents based on paleomagnetism, matching of rock types, mountain ranges, great faults, and orientation of continents to mid-oceanic spreading centers. The early attempts at continental reconstruction were merely schematic, but in 1958 S. Warren Carey produced the first accurate one. Carey's reconstruction is largely in agreement with the most recent computer fitting of the continents done in 1965 by Sir Edward Bullard, J. Everett, and A. G. Smith. More recently, Walter P. Sproll did a computer fit of North America against northeastern Africa.

In 1970 Robert Dietz and John Holden, drawing upon these earlier works, used a reconstruction model in which all of the continents are united into a single Pangaea (see fig. 8.8). The continents are joined at a depth of 6,000 feet, about halfway down the continental slope, since they believe this to be about half the height of the vertical cliffs formed when the continents were pulled apart. The vertical walls formed in separation are thought to have slumped to their present, more gentle, stable slope. For the first time, Dietz and Holden used map coordinates to position the continents in absolute terms during their drift through time. Such unique reconstructions were largely implemented by an idea proposed first by Tuzo Wilson, who believed that hot spots or plumes rise from deep within the mantle and intermittently expel lava through openings onto the sea floor. As the lithospheric plates drift over the hot spots, the outpouring lava forms a line of volcanoes which are older with increasing distance from the hot spot. The line of volcanoes indicates the absolute direction of plate movement past the hot spot. Such a line of volcanic cones extending outward across the ocean floor from the hot spot is called a **thread ridge** or **nematath.** In a series of six world maps extending from 225 million years ago at the end of the Permian period to 50 million years into the future, Dietz and Holden show the breakup of Pangaea and the dispersal of its fragments to form our modern continents.

Projecting present plate movements 50 million years into the future brings Australia up against the Eurasian plate; destruction of the Mediterranean occurs by the northward movement of Africa; and Baja and a slice of western California break off and move northward to be consumed in the Aleutian trench. The Pacific Ocean becomes greatly reduced in size. W. J. Morgan hypothesizes that a hot spot beneath the Yellowstone region "foretells" the breakup of North America.

This book took its start in the notion that the discoveries of radioactive decay and the seismograph around the turn of the century led to the destruction of geology's old unifying theory of a cooling, shrinking earth. At about that time, a poorly nourished and slowly growing embryonic theory of continental drift appeared. In the past two decades, the concept of "drift" has undergone accelerated growth and acquired a new body of support and scientific respectability which amounts to a revolution in the earth sciences. The emergent, more complete theory, renamed **plate tectonics,** appears to be geology's new unifying concept. It holds promise of growth to limits which cannot yet be foretold.

Glossary

Andesite A fine-grained rock intermediate in color and composition between granite and basalt.

Andesite line A map line which separates the island arc and continental regions of andesitic volcanic rocks from the typically basaltic volcanic rocks of the ocean areas.

Asthenosphere A zone of undefined thickness the top of which lies 40 to 90 miles below the earth's surface; in this region of the upper mantle, just below the lithosphere, temperature and pressure reduce the strength of rocks such that they flow plastically.

Batholith A large (more than 40 square miles of surface exposure), shield-shaped, intrusive igneous rock mass generally granitic in composition. The batholith has no discernible floor or regular base.

Benioff (seismic) zone A plane of earthquake activity which slopes about 45°, extending from a deep-sea trench under the adjacent continent. It is theoretically held that lithospheric plates plunge into the upper mantle through this zone and cause earthquakes.

Continental drift The concept that all of the present continents were formerly united into larger landmasses which split apart; the pieces then drifted to assume their present shapes and positions.

Continental nucleus See **Craton.**

Convection current Circulation of material caused by density differences in which hotter, less dense material rises and displaces cooler, denser material. A pot of boiling soup shows such circulation. It is theorized that there is slow (few inches per year), mass movement of mantle material in such convection cells resulting from heat variations in the earth's interior. Convection currents are favored as a possible cause of tectonic plate movement.

Continental rise The sea-floor region between the continental slope and the deep ocean floor; it occurs about 4500 to 17,000 feet and extends down to the abyssal sea floor.

Continental shelf The edge of the continent extending in a gentle slope from the shore to a sharp increase in bottom slope which is the beginning of the continental slope. If there is no break in slope, the bottom limit is placed at 600 feet (200 meters).

Continental slope That part of the edge of the continent which lies between the continental shelf and the continental rise or oceanic trench.

Coral An attached or fixed, bottom-dwelling marine animal which resembles an inverted jellyfish living in a hard limy cup which it secretes. Most grow in colonies, but some are solitary. They have a diverse and an abundant fossil record and are still common in the oceans.

Coral reef A mound or ridge, often very extensive, formed from the growth and death of corals and other marine organisms. Coral reefs generally require a minimum temperature of 18°C for growth and thus are restricted to lower latitudes.

Core of the earth The innermost or central zone of the earth which is divided into an inner core (solid) and an outer core (liquid). The core is probably composed of an iron-nickel alloy.

Correlation A demonstration that two separated rock units were formed at the same time. Such equivalency in age is determined by index fossils or measurement of proportions of certain chemical elements which decay radioactively to yield new elements.

Cosmic ray Very high-speed subatomic particles which bombard the earth from outer space.

Craton The central part of continents which have been stable for great periods of geologic time. They are composed mainly of highly metamorphosed, very ancient rocks (Precambrian).

Cretaceous A period of the geologic time scale which lasted from about 135 to 63 million years ago.

Crustal plate See **Lithospheric plate.**

Crust of the earth The uppermost layer or shell of the earth which, strictly defined, lies above the Moho, or Mohorovčić discontinuity. **Crust** is used in imprecise senses and is often applied to the lithosphere.

Daughter element An element formed from another element (parent) through radioactive decay.

Dike A tabular or sheet-like body of igneous rock that cuts across the structure or layering of the rock mass which it intrudes or in which it forms.

Dunite A coarse-grained igneous rock composed almost completely of the green mineral olivine; dunite appears to have been thrust to the earth's surface from its place of origin in the mantle. It may represent upper-mantle material.

Earthquake Shaking or vibration of the earth produced by sudden movements in the earth's interior.

Eclogite A rare metamorphic rock that forms under very high pressure. It is similar in composition to the common igneous rock called basalt. It is composed almost entirely of grains of the minerals garnet and pyroxene.

Endemic organisms Organisms which are limited in distribution to a specific area or environment. They are indigenous, native, or peculiar to a particular region.

Eocene An epoch of the Tertiary period of the geological time scale. It lasted from about 58 to 36 million years ago.

Epicenter A point on the earth's surface which lies directly above the true center or focus of an earthquake.

Epoch A subdivision of a period in the geologic time scale.

Era A major subdivision of the geologic time scale.

Eugeocline (Eugeosyncline) The part of a geosynclinal couplet which lies seaward of the continental slope. It lies deeper than and beyond the migeocline. (See **Geosynclinal couplet** and **Geosyncline.**) Clastic sediments are associated with volcanic products.

Extrusive igneous rock A rock which solidifies from molten lava on the earth's surface. A volcanic rock.

Fault A fracture surface or zone of breakage in the earth along which one rock mass moves relative to that on the other side.

Fauna The animals of a given area or time that lived in association with each other. We speak of the fauna of North America or the fauna of the Cretaceous period or the mammalian fauna of the Cretaceous of North America.

Faunal succession, principle of William Smith's contribution of 1815 that certain fossil species occur in ordered sequences in sedimentary rocks such that fossils are a guide to relative age of the rock bed in which they are found.

Flora The complete plant population of a given area, formation, environment, or interval of geologic time, e.g., the flora of the Arctic; the flora of the Pennsylvanian period; the flora of the Cenozoic of Africa.

Focus The point of origin of an earthquake within the earth, at which stored strain energy is suddenly converted to shock wave energy.

Foraminifera Shelled, single-celled, usually microscopic animals which are valuable in dating rocks of several periods. Most commonly found in marine and brackish water, but also known to live in freshwater. They range from the Precambrian to the present.

Formation (rock) The basic unit into which rock beds are grouped for descriptive purposes and geologic mapping. Most formations have distinctive rock features which allow their recognition in the field.

Fossil Any trace or evidence of prehistoric life.

Genetic Those characteristics of an organism due to the action of genes or inheritance from the preceding generation. Genetic effects are distinguished from environmental effects.

Genus (plural, Genera) A unit of biological classification composed of species similar in structure and ancestry; the genus may contain only a single species. The genus ranks below and is grouped with other genera into families.

Geosynclinal couplet (Also see **Geosyncline.**) The classic geosyncline is divided into two parts or a couplet. The eugeocline (lying farther offshore) is adjacent and parallel to the miogeocline which lies against the continent. A modern example of such a couplet may be the wedge of sediments which caps the continental slope off the U.S. east coast; this area appears to be a miogeocline. The deeper-water continental rise lying seaward is seemingly a eugeocline in which the characteristic turbidite sediments (graywacke) accumulate.

Geosyncline A large, elongate, gradually subsiding trough or basin in which a thick succession of sedimentary beds accumulates (thousands of feet) over a great period of time (several geologic periods). Geosynclines generally form along the edges of continents and are destroyed during episodes of crustal deformation, often producing folded mountain ranges such as the North American Appalachians. Many different kinds of geosynclines have been defined. The classic geosyncline is broken into two adjacent and parallel structures: the eugeosyncline (true geosyncline) and the miogeosyncline (lesser geosyncline).

Geothermal gradient The gradual and steady increase of temperature with increasing depth within the earth, often expressed as degrees of temperature per unit depth.

Glacier A mass of moving ice formed from accumulated snow which is compacted and recrystalized. It moves by the stress of its own weight. Alpine or mountain glaciers occupy smaller areas than continental ice sheets or caps which cover vast expanses up to continental size.

Global tectonics (New global tectonics) A term often used as a synonym for **Plate tectonics.** Relates to the processes of crustal movement and deformation of lithospheric plates. Tectonics on a worldwide scale involving very large-scale movements within the earth. (See **Plate tectonics.**)

Glossopteris flora A fossil plant assemblage characterized by the genera **Glossopteris** and **Gangamopteris.** The flora is found in the Gondwanaland continents of the Southern Hemisphere; it is thought to have been composed mostly of low shrubs which grew in cool climates near the late Paleozoic glaciers and were deposited during the interglacials to form coal beds. The flora is important evidence for continental drift.

Gondwana (Gondwanaland) The southern protocontinent believed to have included Antarctica, Africa, South America, Australia, and India. It is named for the Gondwana system (rock beds) of India which are late Paleozoic and early Mesozoic in age, and which contain characteristic glacial tillites below coal measures. Similar rock sequences of the same age containing identical fossil plants **(Glossopteris flora)** suggested former connection of the southern continents. The northern hypothetical supercontinent was Laurasia; both Gondwana and Laurasia are thought to have been derived by the splitting of Pangaea.

Granite A rock formed from the cooling of molten rock (magma) which develops crystal grains large enough to be seen with the naked eye. The minerals quartz and feldspar are dominant. Granitic igneous rocks are mainly confined to the continents; in contrast, basalts underlie the ocean floors.

Gravity anomaly A gravity measurement value that is greater (positive anomaly) or smaller (negative anomaly) than the predicted or expected value at a particular place.

Graywacke; greywacke A type of sandstone that is gray or greenish-gray, hard and tough, and composed of angular grains of quartz, feldspar, and rock fragments held together in a clayey matrix. Normally associated with submarine lava flows and cherts, they are thought to be deposited by submarine turbidity currents.

Guide fossils Fossil species useful in the relative dating of rocks. Guide fossils have the characteristics of short species life, widespread geographic distribution, abundance, and easy recognition.

Half-life The time required for one-half of an original number of parent atoms to decay to yield daughter atoms.

Hot spot (thermal center) Place at which volcanic outpourings occur as a result of rising columns or plumes from deep within the mantle. Such plumes or thermal centers are thought to remain fixed in position for 100 million or more years, producing lines of volcanoes (nemataths) as tectonic plates drift over them. Groups of hot spots have been suggested as causes of plate motions.

Intrusive rock An igneous rock which solidifies from magma (molten rock) beneath the earth's surface. Such rocks contain large mineral grains; in contrast, extrusive or volcanic rocks contain small grains or form glasses.

Island arc A chain of islands rising from the ocean floor which are arranged in an arc shape with the convex side of the curve facing the open ocean. The arc lies near a continent with a deep trench bordering it on the ocean side and a deep-sea basin between the arc and continent. Island arcs are associated with deep-focus earthquakes and negative gravity anomalies.

Isostasy The theory that the large blocks of the earth's crust float on the substratum in the mantle; that the crust tends to attain a state of floating balance with respect to the material beneath it.

Laurasia The great northern protocontinent from which the present continents of the Northern Hemisphere have been derived by splitting and continental drift. Laurasia, together with Gondwana, the southern protocontinent, comprised Pangaea, the original, single earth landmass. Laurasia included North America, Eurasia exclusive of India, and Greenland. The main line of breakup lay in the North Atlantic with a split extending into Hudson Bay.

Lithology The description of rocks especially as based on studies of hand specimens and outcrops. The physical character of a rock including composition, grain size, color, structure, etc.

Lithosphere The solid outer shell of the earth which varies from 40 to 90 miles thick. It is composed of the earth's crust and the uppermost rigid layer of the mantle. The lithosphere lies above the asthenosphere and comprises the rigid plates which move in plate tectonics.

Lithospheric plate A segment or slab of brittle, rigid lithosphere composed of crust and uppermost mantle material. The earth's outer shell is composed of eight large and about two dozen smaller lithospheric plates which move over the plastic asthenosphere below.

Magnetic polar reversals An as yet unexplained but definitely verified phenomenon in which the north magnetic pole becomes the south magnetic pole and vice versa. It has now been shown that the earth's magnetic field has two stable states; the magnetic field has reversed itself repeatedly between them. A time scale of magnetic reversals has been drawn for the past several million years.

Magnetic poles The two areas at opposite ends of a magnet which show greatest magnetic force. The lines of magnetic force leave the magnet at the north or positive pole and bend to enter the south or negative pole.

Mantle The interior earth zone between the core and the crust; it is divided into the upper mantle, which shows a rapid increase in density with depth, and the lower mantle, which gradually becomes denser with depth.

Metamorphism The process which involves the application of heat, pressure, and the introduction of new chemical substances to rocks; as a result the rocks are altered in composition, texture, and/or structure. Usually new minerals develop during metamorphism.

Miogeocline (miogeosyncline) The shallower part of a geosynclinal couplet which lies between the continent and the deeper, offshore eugeocline. (See **Geosyncline** and **Geosynclinal couplet**.)

Nematath See **Thread ridge.**

Nucleus See **Craton.**

Ocean-floor spreading See **Sea-floor spreading.**

Ophiolites Masses of very basic igneous rocks such as basalt, gabbro, and olivine-rich rocks which have apparently been moved as wedges and slivers; their structure and composition suggests that they may be pieces of uppermost mantle or ocean-floor crust which broke from descending plates and were thrust upward into overlying rocks. Ophiolites therefore are believed to lie along the sites of former subduction zones and mark the line along which continents collided to form mountain ranges.

Orogeny The process of mountain building especially with reference to folding and faulting.

Paleobiogeography The science which treats the geographic distribution of living organisms during ancient times.

Paleoecology The study of the relationships between organisms and their environments during the geologic past. Death, burial, and postburial interpretations are based on fossil plants and animals and the nature and history of the rock in which they are entombed.

Pangaea A supercontinent believed to have existed early in geologic history, comprised of all of the present continents before their fragmentation and drift. At some stage of breakup Pangaea is thought to have split into two large pieces, Gondwana to the south and Laurasia on the north.

Peridotite A general term for a coarse-grained, intrusive igneous rock composed mainly of the green mineral olivine; it may or may not contain other iron-magnesium-bearing minerals such as amphiboles or pyroxenes and lacks feldspar. The upper mantle may be composed of peridotite.

Plate tectonics **Tectonic** refers to the deformation of the earth's lithosphere and the rock structures and surface features produced by such movement. The theory of plate tectonics includes a model of the earth with a rigid outer shell or lithosphere divided into about 25 huge slabs or plates which ride or float on a viscous zone beneath them in the upper mantle (asthenosphere). The plates move about independently pushed outward from **spreading centers** (mid-oceanic ridges) where the plates grow; the plate edges opposite the spreading centers dive into **subduction zones** to be consumed in the mantle. The third kind of plate edge is a **transform fault** along which plates slide past each other. The continents are merely passengers riding the plates. The major mountain belts and earthquake zones occur at plate boundaries.

Sea-floor spreading The concept that the oceanic ridges are axes along which new ocean crust is generated continually by injection of rock from the mantle. The ocean floor moves away from the ridges and toward subduction zones into which the leading edge of the crust plunges to be consumed in the mantle.

Serpentine A common, rock-forming mineral group which has a greasy luster, soapy feel, and is commonly green, greenish-gray, greenish-yellow or, less often, brown, black, or white. Serpentines are formed by the metamorphism or alteration of magnesium silicate minerals such as olivine.

Silica Silicon dioxide, SiO_2; an abundant and chemically resistant compound which occurs as the mineral quartz, opal, chert, or others. It is combined as an important constituent in the common, rock-forming silicate materials such as feldspar, olivine, augite, hornblende, etc.

Spreading center A mid-oceanic ridge along which tectonic plates separate and drift apart. The spreading of plates is accompanied by upwellings of lava which fills the tensional cracks of the spreading center (rift zone) and cools to form new basaltic crust. Thus new surface area is created at spreading centers.

Spreading pole An axis or a pole of rotation around which points lying on two different tectonic plates may be moved by rotation of the rigid plates. Minimum movement occurs at the rotational equator which lies 90° away from the spreading pole.

Stratigraphy The branch of geology which deals with the description, definition, sequence, correlation, and formation of the layered rocks of the earth's crust.

Subduction zone A plate boundary along which two lithospheric plates converge or collide and the leading edge of one is thrust or dives under the other. The subducted or sinking plate is transformed into mantle material.

System Those rocks formed during an interval of geologic time called a **period,** e.g., the Permian system. A major time-stratigraphic unit defined with a designated type area and correlated with other regions mainly by the contained fossils.

Tethys Sea (Tethyian geosyncline) A sea of which the present Mediterranean is thought to be a vestige. Tethys lay between the northern and southern continents of the Eastern Hemisphere from the Permian through the early Tertiary periods along the region of the Alpine-Himalayan mountain belt. Laurasia and Gondwanaland were thought to have been separated by Tethys.

Thermal center See **Hot spot.**

Thread ridge (nematath) A line of volcanoes which forms as a lithospheric plate drifts over a hot spot. Intermittent volcanic outpourings form a line of volcanoes which are older with increasing distance from the hot spot. The direction of the line of volcanoes indicates the absolute direction of plate movement past the hot spot.

Tillite A sedimentary rock formed by the compaction and cementation of glacially deposited, unsorted sediment called **till.**

Transform fault (Transform) An earthquake, active, strike-slip fault which ends at mid-oceanic ridge crests or subduction zones. Transforms are not simple strike-slip faults; movement is now active only in the segment of the fault which lies between a pair of ridges where new crust forms or between a pair of subduction zones where crust descends into the mantle. The sense of movement on transforms is actually the reverse of what it appears to be.

Triple junction A place at which three tectonic plates meet. There are stable triple junctions which keep their geometry as plates drift and unstable ones which change with drift. Six different triple junctions composed of various combinations of intersections of spreading centers, subduction zones, and transforms are found at present, but sixteen types are theoretically possible.

Uniformitarianism The doctrine or principle that the natural laws and processes operating on and within the earth at present have acted in much the same way and intensity throughout geologic time. The idea that all of the geologic features of the earth have been produced by forces and events which are still observable today. "The present is the key to the past."

Volcanic island arc See **Island arc.**

Supplementary Reading

References particularly suitable for the introductory student are preceded by an asterisk. Because the subject matter is new and rapidly evolving, the reader may wish to turn to recent issues of **Scientific American, Nature, Science Digest, The American Scientist, New Scientist, Science,** etc.

Chapter 1 — The Crust and Interior of the Earth

*Anderson, D. L. "The Plastic Layer of the Earth." **Scientific American** 207 (July 1962): 52–59.

*Bullen, K. E. "The Deep Interior." **The Earth and Its Atmosphere.** Edited by D. R. Bates. New York: Science Editions, Inc., 1961, pp. 31–47 (paperback).

*Hodgson, J. H. **Earthquakes and Earth Structure,** Englewood Cliffs. Prentice-Hall, Inc., 1964 (paperback).

Robertson, E. C., ed. **The Nature of the Solid Earth.** New York: McGraw-Hill, 1972.

*Takahashi, Taro, and Basset, W. A. "The Composition of the Earth's Interior," **Scientific American** 212 (June 1965): 100–108.

*Tocher, Don. "Earthquakes and Rating Scales." **Geotimes** 8 (May–June 1964): 15–18.

Chapter 2 — Geologic Time and Dating Rocks

Berry, W. B. N. **Growth of a Prehistoric Time Scale Based on Organic Evolution.** San Francisco: W. H. Freeman & Co., 1968, pp. 1–136.

*Eicher, D. L. **Geologic Time.** Englewood Cliffs: Prentice-Hall, Inc., 1968 (paperback).

*Donovan, D. T. **Stratigraphy, Introduction to Principles.** Chicago: Rand McNally, 1966.

Kulp, J. L. "Geologic Time Scale." **Science** 133 (1961): 1105–14.

*Raup, D. M., and Stanley, S. M. **Principles of Paleontology.** San Francisco: W. H. Freeman and Co., 1971.

Chapter 3 — Early Evidence for Continental Drift

Bullard, E.; Everett, J. E.; and Smith, A. G. "The Fit of the Continents Around the Atlantic." **Philosophical Transactions of the Royal Society of London** 258 (1965): 41–51.

Carey, S. Warren. "The Asymmetry of the Earth." **Australian Journal of Science** 25 (1963): 369–83, 479–88.

Du Toit, A. L. **Our Wandering Continents.** 1937. Reprint. Edinburgh: Oliver and Boyd Ltd., 1957.

Ewing, J., and Ewing, M. "Sediment Distribution on the Mid-Ocean Ridges with Respect to Spreading of the Sea Floor." **Science** 156 (1967): 1590.

Martin, H. "The Hypothesis of Continental Drift in the Light of Recent Advances of Geological Knowledge in Brazil and in South West Africa." Alexander L. Du Toit Memorial Lecture (1961) no. 7, **Transactions of the Geological Society of South Africa,** Annexture to vol. 64, pp. 1–47.

*Wegener, A. **The Origin of Continents and Oceans.** New York: Dover Publications, Inc., 1966 (paperback).

Chapter 4 — Fossil Life and Climates of the Past

Bain, G. W. "Climatic Zones Throughout the Ages." In **Polar Wandering and Continental Drift,** edited by A. C. Munyan. Tulsa: Society of Economic Paleontologists and Mineralologists, 1963, pp. 94–130.

Brown, D. A. "Diversity of Triassic Reptiles Plotted Against Present and Triassic Latitudes Based on Paleomagnetism." **Australian Journal of Science** 30 (1968): 440.

*Clark, D. L. **Fossils, Paleontology and Evolution.** Dubuque, Iowa:. W. C. Brown, 1968 (paperback).

*Colbert, E. H. **Evolution of the Vertebrates.** New York: John Wiley and Sons, 1969.

Crowell, J. C., and Frakes, L. A. "Phanerozoic Glaciation and the Causes of the Ice Ages." **American Journal of Science** 268 (1970): 193–224.

*Doumani, G. A., and Long, W. E. "The Ancient Life of the Antarctic." **Scientific American** 207 (1962): 169–84.

*Hallam, A. "Continental Drift and the Fossil Record." **Scientific American** 227 (November 1972): 55–66.

Hays, J. D. "Faunal Extinctions and Reversals of the Earth's Magnetic Field." **Bulletin of Geological Society of America** 82 (1971): 2433–47.

*Kurtén, Björn. "Continental Drift and Evolution." **Scientific American** 220 (March 1969): 54–64.

McKenna, M. C. "Sweepstakes, Filters, Corridors, Noah's Arks, and Beached Viking Funeral Ships in Paleogeography." **Implications of Continental Drift to the Earth Sciences**, edited by D. H. Tarling and S. K. Runcorn, vol. 1. London and New York: Academic Press, 1973, pp. 295–308.

*——— "Possible Biological Consequences of Plate Tectonics." **Bioscience** 22 (September 1972): 519–25.

*Newell, N. D. "Crises in the History of Life." **Scientific American** 208 (February 1963): 76–92.
*——— "The Evolution of Reefs." **Scientific American** 226 (June 1972): 54–65.
Opdyke, N. D. "Paleoclimatology and Continental Drift." In **Continental Drift.** International Geophysics Series, no. 3. pp. 41–66, 1962.
*Simpson, G. C. **Life of the Past.** New Haven, Conn.: Yale University Press, 1961 (paperback).
*Valentine, J. W., and Moores, E. M. "Plate Tectonics Regulation of Faunal Diversity and Sea Level: A Model." **Nature** 228 (1970): 657–59.
*Valentine, J. W., and Eldridge, M. M. "Global Tectonics and the Fossil Record." **Journal of Geology** 80 (March 1972): 167–84.

Chapter 5 — Rising and Sinking Crust, Heat from the Earth, and Currents in the Mantle

*Cailleux, A. **Anatomy of the Earth.** New York: McGraw-Hill, 1968 (paperback).
*Elsasser, W. M. "The Earth as a Dynamo." **Scientific American** 198 (May 1958): 44–48.
——— "Convection and Stress Propagation in the Upper Mantle." **New Castle Symposium.** New York: John Wiley and Sons, 1968.
*Heiskanen, W. A. "The Earth's Gravity." **Scientific American** 193 (September 1955): 164–74.
Lanseth, M. G., Jr.; and Von Herzen, R. P. "Heat Flow Through the Floor of the World Oceans." **The Sea** 4 (1970) Interscience.
Lubimova, E. A. "Thermal History of the Earth." Monograph 13 on **The Earth's Crust and Upper Mantle.** American Geophysical Union, 1969.
Roy, R. F.; Blackwell, D. D.; and Decker, E. R. "Continental Heat Flow." **The Nature of the Solid Earth.** New York: McGraw-Hill, 1972.
*Verhoogen, John. "Temperatures Within the Earth." **American Scientist** 48 (June 1960): 134–59.

Chapter 6 — Fossil Magnetism in Rocks

Blackett, P. M. S. "Comparisons of Ancient Climates with the Ancient Latitudes Deduced from Rock Magnetic Measurements." **Proceedings of the Royal Society** A263 (1961): 1–30.
Creer, K. M. "A Review of Paleomagnetism." **Earth Science Reviews** 6 (1970): 369–466.
*Cox, Allan; Dalrymple, G. B.; and Doell, R. R. "Reversals of the Earth's Magnetic Field." **Scientific American** 216 (February 1967): 44–54.
Cox, Allan, and Doell, R. R. "Review of Paleomagnetism." **Bulletin of Geological Society of America** 71 (June 1960): 645–768.

*Irving, E. **Paleomagnetism and Its Application to Geological and Geophysical Problems.** New York: John Wiley and Sons, 1964. Elmsford, N.Y.:

Jacobs, J. A. **Earth's Core and Geomagnetism.** Elmsford, N.Y.: Pergamon Press, Inc., 1963, (paperback).

Opdyke, N. D. "Paleomagnetism of Deep-Sea Cores." **Reviews of Geophysics and Space Physics** 10 (1972): 213–49.

Chapter 7 — Plate Tectonics

Atwater, T. "Implications of Plate Tectonics for the Cenozoic Tectonics Evolution of Western North America." **Bulletin of Geological Society of America** 81 (1970): 3513–36.

Beloussov, V. V. "Against the Hypothesis of Ocean Floor Spreading." **Tectonophysics** 9 (1970): 489–511.

*Bullard, Sir Edward. "The Origin of the Oceans." **Scientific American** 221 (September 1969): 66–75.

Dewey, J. F., and Bird, J. M. "Mountain Belts and the New Global Tectonics." **Journal of Geophysical Research** 75 (May 1970): 2625–45.

——— "Origin and Emplacement of the Ophiolite Suite: Appalachian Ophiolites in Newfoundland." **Journal of Geophysical Research** 76 (1971): 3179–3206.

*Dewey, J. F. "Plate Tectonics." **Scientific American** 226 (May 1972): 56–68.

Dickinson, W. R. "Relation of Andesites, Granites and Derivative Sandstones to Arc-trench Tectonics." **Review of Geophysics and Space Physics** 8 (1970): 813–60.

——— "Plate Tectonics in Geologic History." **Science** 174 (1971): 107–13.

Dietz, R. S. "Continent and Ocean Basin Evolution by Spreading of the Sea Floor." **Nature** 190 (1961): 854–57.

*———. "Geosynclines, Mountains and Continent Buildings." **Scientific American** 226 (March 1972): 30–38.

*Emery, K. O. "The Continental Shelves." **Scientific American** 221 (September 1969): 107–22.

Ewing, J., and Ewing, M. "Sediment Distribution of the Mid-Ocean Ridges with Respect to Spreading of the Sea Floor." **Science** 156 (1967): 1590–92.

*Fisher, R. L., and Reville, R. "The Trenches of the Pacific." **Scientific American** 193 (November 1955): 36–41.

*Heezen, B. C. "The Rift in the Ocean Floor." **Scientific American** 203 (October 1960): 98–110.

*Heirtzler, J. R. "Sea Floor Spreading." **Scientific American** 219 (1968): 60–70.

Hess, H. H. "History of Ocean Basins." **Petrologic Studies.** pp. 599–620. Volume to honor A. F. Buddington, edited by A. E. J. Engel, et al. New York: Geological Society of America, 1962.

*Hurley, P. M. "The Confirmation of Continental Drift." **Scientific American** 218 (1968): 52–64.

Hurley, P. M., and Rand, J. R. "Pre-drift Continental Nuclei." **Science** 164 (1969): 1229–42.

Isacks, B. I.; Oliver, J.; and Sykes, L. R. "Seismology and the New Global Tectonics." **Journal of Geophysical Research** 73 (1968): 5855–99.

Le Pichon, X. "Sea-floor Spreading and Continental Drift." **Journal of Geophysical Research** 73 (1968): 3661–97.

Maxwell, A. E., ed. **The Sea.** New Concepts of Sea Floor Evolution, vol. 4. New York: Wiley-Interscience, 1970.

McKenzie, D. P., and Morgan, W. J. "The Evolution of Triple Junctions." **Nature** 224 (1969): 125–33.

*Menard, H. W. "The Deep-Ocean Floor." **Scientific American** 221 (1969): 126–42.

Mitchell, A. H., and Reading, H. G. "Continental Margins, Geosynclines, and Ocean Floor Spreading." **Journal of Geology** 77 (1969): 629–46.

Morgan, W. J. "Rises, Trenches, Great Faults, and Crustal Blocks." **Journal of Geophysical Research** 73 (1968): 1959–82.

*Orowan, Egon "The Origin of the Oceanic Ridges." **Scientific American** (November 1969): 102–19.

Vine, F. J. "Spreading of the Ocean Floor: New Evidence," **Science** 154 (1966): 1405–15.

*———. "Sea Floor Spreading — New Evidence." **Journal of Geological Education** 17 (February 1969): 6–16.

Vine, F. J., and Matthews, D. H. "Magnetic Anomalies Over Oceanic Ridges." **Nature** 199 (1963): 947–49.

*Wilson, J. T. "Continental Drift." **Scientific American** 208 (1963): 86–100.

Wilson, J. T. "A New Class of Faults." **Nature** 207 (1965): 343–47. Reprinted in **Adventures in Earth History,** p. 351.

———. "Did the Atlantic Close and Then Reopen?" Nature 211 (1966): 676–81.

Chapter 8 — Plates Past, Present, and Future: A Schedule for Drifting

Dietz, R. S., and Holden, J. C. "Reconstruction of Pangaea: Breakup and Dispersion of Continents, Permian to Present." **Journal of Geophysical Research** 75 (1970): 4939–56.

*———. "The Breakup of Pangaea." **Scientific American** (October 1970): 30–41.

Gilluly, J. "Tectonics Involved in the Evolution of Mountain Ranges." In **The Nature of Solid Earth,** Birch Symposium. New York: McGraw-Hill, 1971.

Lyustikh, E. H. "Criticism of Hypothesis of Convection and Continental Drift." **Geophysical Journal of the Royal Society** 14 (1967): 347–52.

Meyerhoff, A. A., and Meyerhoff, H. A. "The New Global Tectonics: Major Inconsistencies." **Bulletin of the American Association of Petroleum Geologists** 56 (1972): 269–336.

———. "The New Global Tectonics: Age of Linear Magnetic

Anomalies of Ocean Basins." **Bulletin of the American Association of Petroleum Geologists** 56 (1972): 337–59.

Morgan, W. J. "Deep Mantle Convection Plumes and Plate Motions." **Bulletin of the American Association of Petroleum Geologists** 56 (1972): 203–13.

*Rona, P. A. "Plate Tectonics and Mineral Resources." **Scientific American** (July 1973): 86–95.

*Tazieff, Haroun. "The Afar Triangle." **Scientific American** 222 (February 1970): 32–40.

Wessen, P. S. "Objections to Continental Drift and Plate Tectonics." **Journal of Geology** 80 (1972): 185–97.

Chapter 9 — Historical Summary

*Adams, F. D. **The Birth of the Geological Sciences.** New York: Dover Publications, Inc., 1954 (paperback).

Carozzi, A. V. "New Historical Data on the Origin of the Theory of Continental Drift." **Geological Society of America Bulletin** 81 (1970): 283–86.

Geikie, Sir Archibald. **The Founders of Geology.** New York: Dover Publications, Inc., 1962.

Georgi, J. "Memories of Alfred Wegener." In **Continental Drift.** Edited by S. K. Runcorn. New York: Academic Press, 1962.

Marvin, U. B. **Continental Drift — The Evolution of a Concept.** Wash., D. C.: Smithsonian Institution Press, 1973.

Rupke, N. A. "Continental Drift Before 1900." **Nature** 227 (1970): 349–50.

Snider-Pellegrini, A. **La Création et ses Mystères Devoilés.** Paris: Franck & Dentu, 1858.

Taylor, F. B. "Bearing of the Tertiary Mountain Belt on the Origin of the Earth's Plan." **Bulletin of the Geological Society of America** 21 (1910): 179–226.

*Van Waterschoot van der Gract et al. "Theory of Continental Drift." Tulsa: American Association of Petroleum Geologists, 1928.

Wegener, A., Petermanns Mitt. 58, 185, 253, 305, (1912). In book form: Die Entstehung der Kontinente und Ozeane, (Sommlung Vieweg, H. 23), Vieweg and Son, Brunswick (1915): 2nd. Ed. 1920: 3rd. Ed. 1922: 4th Ed. 1924, revised by A. Wegener, 1929: 5th Ed. revised by Kurt Wegener, 1936: 3rd Ed. translated into English, French, Spanish, Swedish and Russian. 4th Ed. reprinted 1961, (Sammlung Die Wissenschaft). English translation of the 1929 Ed., New York, Dover, 1966, 246 pp. (paperback).

Wegener, Else, ed., assisted by Fritz Loewe. **Greenland Journey.** Translated from the 7th German ed. by Winifred M. Dean. London: Blackies and Son, Ltd., 1939.

Wilson, J. T., and Beloussov, V. V. "Debate About the Earth." **Bulletin of the Canadian Institute of Mining and Metallurgy** (1968). Reprinted with the pro and con open letters by Wilson & Beloussov in **Geotimes,** Dec. 1968, and in **The Nature of Geology,** Brainard Mears, Jr. City. Van Nostrand Reinhold Co., 1970

General Comprehensive Works

*Cloud, Preston. **Adventures in Earth History.** San Francisco: W. H. Freeman and Co., 1970. A collection of papers by many authors.

Cox, Allan, ed. **Plate Tectonics and Geomagnetic Reversals.** San Francisco W. H. Freeman and Co., 1973.

Foster, R. J. **General Geology.** 2d ed. Columbus, Ohio: Charles E. Merrill Publishing Co., 1973.

Gass, I. G. et al., ed. **Understanding the Earth.** Published for the Open University. Sussex: Artemus Press, 1971.

*Hallam, A. "A Revolution in the Earth Sciences." **Continental Drift to Plate Tectonics.** Oxford and London: Clarendon Press, 1973.

*Holmes, A. **Principles of Physical Geology,** 2d ed, New York: Ronald Press, 1965.

*Kummel, Bernhard. **History of the Earth.** San Francisco: W. H. Freeman and Co., 1961.

*Mintz, L. W. **Historical Geology.** Columbus, Ohio: Charles E. Merrill Publishing Co., 1972.

*Takeuchi, H.; Uyeda, S.; and Kanamori, H. **Debate About the Earth.** San Francisco: Freeman Cooper and Co., 1967 (paperback).

*Tarling, D., and Tarling, M. **Continental Drift, A Study of the Earth's Moving Surface.** New York: Anchor Books, Doubleday and Co., Inc., 1971.

*Tarling, D. H., and Runcorn, S. K., ed. **Implications of Continental Drift to the Earth Sciences,** Vol. 1: (1) Paleomagnetic Review; (2) Sea-Floor Spreading Review; (3) Paleontological Implications; (4) Paleoclimatic Implications; (5) Economic Significance. Vol. 2: (1) Rift Margins and Continental Edge Structures; (2) Rifts and Oceans; (3) Paleogeographic Implications; (4) Ancient Oceans; (5) Continental Evolution; (6) Epilogue. New York: Academic Press, Inc., 1972 and 1973.

*Wessen, P. S. "Objections to Continental Drift and Plate Tectonics." **Journal of Geology** 80 (March 1972): 185–97.

*Wilson, J. T. "Continents Adrift." **Scientific American**. San Francisco: W. H. Freeman and Co., 1970 (paperbound). An excellent collection of fifteen short papers covering many aspects of drift.

Index